Anonymous

The natural wonders of New Zealand

Its boiling lakes, steam holes, mud volcanoes, sulphur baths, medicinal springs, and burning mountains

Anonymous

The natural wonders of New Zealand
Its boiling lakes, steam holes, mud volcanoes, sulphur baths, medicinal springs, and burning mountains

ISBN/EAN: 9783743466944

Manufactured in Europe, USA, Canada, Australia, Japa

Cover: Foto ©berggeist007 / pixelio.de

Manufactured and distributed by brebook publishing software (www.brebook.com)

Anonymous

The natural wonders of New Zealand

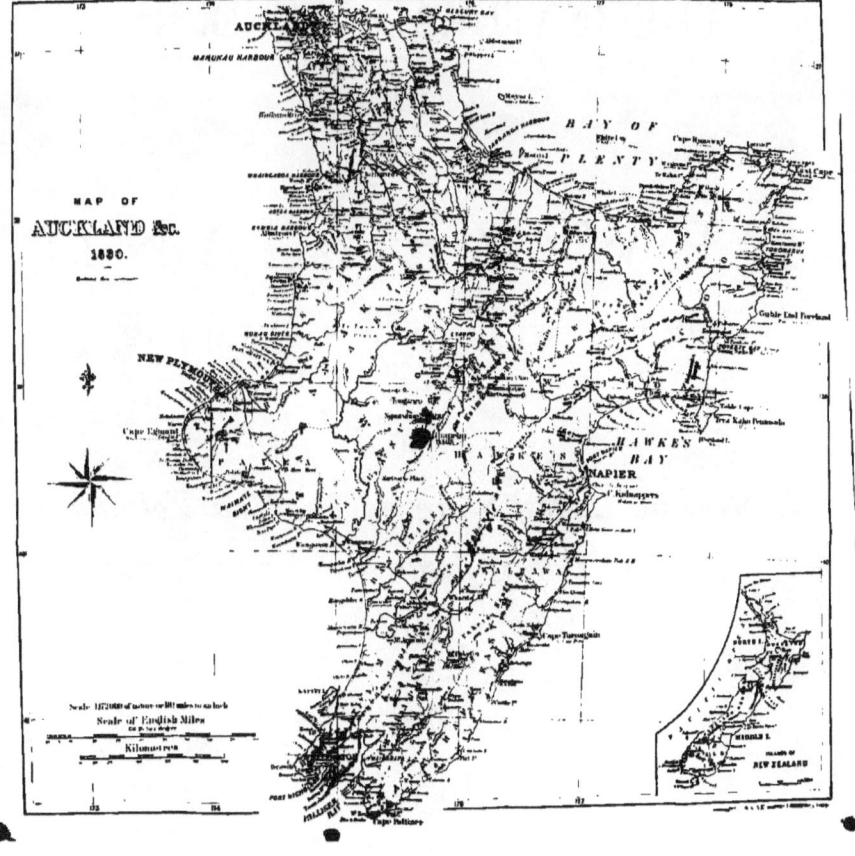

THE NATURAL WONDERS

OF

NEW ZEALAND

(THE WONDERLAND OF THE PACIFIC):

Its Boiling Lakes, Steam Holes, Mud Volcanoes Sulphur Baths, Medicinal Springs, and Burning Mountains.

SECOND EDITION.

LONDON:
EDWARD STANFORD, 55 CHARING CROSS.
1881.

PREFACE.

To give a popular description of the marvellous natural phenomena of boiling hot wells and rivers, mud and lakes, steam-holes, and numerous valuable medicinal springs and baths of hot and cold water and hot mud in the neighbourhood of Auckland, is the object of the present work.

Our first edition was imperfect, it being then almost impossible for a European, in the troubled state the natives were in, to visit and examine carefully the Hot Lake District.

The Government have now acquired, by purchase, some of the most important blocks of land, including several hot lakes and medicinal springs, and are negotiating for the purchase of the whole Lake District, to be reserved and set apart as a great national park for the colony.

The old Maoris cling tenaciously to their gardens, villages, and the sacred burial grounds of their ancestors; and although these are to be reserved for their exclusive use, they dread the influx of the pale-face (Pakeha), who has little respect or sympathy for the associations connected with their family lands.

The young natives are eager to sell or lease these

lands, for they fully appreciate the value of the white man as a neighbour, with his silver and gold, and all the comforts and conveniences these can purchase; but the old chiefs sit listlessly on the hot stones, with their blankets around them, smoking their own tobacco out of their own pipes, while dreaming of the good old times when they were allowed unmolested to enjoy a good feed of baked Pakeha or roasted Maori.

To get to the Hot Lake District with ease and comfort, and escape exorbitant charges, strangers should come to Auckland, where they have daily opportunities by rail and coach *viâ* the Waikato, Ngaruawahia (the ancient capital of the Maori kingdom), Hamilton, and Cambridge (large flourishing country towns), to Ohinemutu on Lake Rotorua.

The weekly mail-coach from Wellington, Napier, and Tauranga has been discontinued by the Government in favour of the more direct route from Auckland by the Waikato, and in future all passengers and stores will be sent by this route, as it is the shortest and most convenient.

Two new hotels are being built at Ohinemutu; the Wairoa hotels are under respectable management; and visitors can now make themselves as comfortable at Tapu-wae-haruru (Taupo) as they could be in Auckland.

Intending visitors can have every information as to charges and requirements, by application at the publishing office of the *New Zealand Almanac*, Queen Street, Auckland.

CONTENTS.

	PAGE
INTRODUCTION,	9
From Auckland by the Waikato to the Hot Lakes,	19
General Cameron's First Engagement in Waikato—the Battle of Koheroa,	22
The Battle of Rangiriri,	24
Auckland to Tauranga—Sea Voyage about Twenty Hours,	28
Tauranga Harbour,	33
Tauranga to Ohinemutu by Stage Coach,	34
The Story of the Massacre at the Gate Pa,	36
The Oropi Eighteen-mile Bush,	42
Tauranga to Maketu, 17 miles; thence to Taheke, on Lake Rotoiti, 21 miles; thence to Ohinemutu, 12 miles,	45
Tikitere Sulphur Springs,	47
Lake Rotorua,	49
Ohinemutu,	51
The Maori Legend of Hinemoa,	59
Sulphur Point,	67
Whaka-rewa-rewa Springs,	69
Ohinemutu to Wairoa, 9 miles,	78
Lake Tikitapu,	78
Rotokakahi,	79
Lake Tarawera,	80
Rotomahana,	85
The White Terrace,	89
The East Side of Lake Rotomahana,	99
The Great Nga-hapu,	99
The Great Steam Fog-horn,	101
Te Takapo Fountains,	102
The Wai-kanapanapa Lower Springs,	102

Contents.

	PAGE
The Kuakiwi, the Ngawhana, the Koingo, and the Whatapohu Fountains,	104
A Fountain, Solfatara, and Fumarole, all three in one,	104
Te Kapiti Fountain,	105
Roto Pounamu, or the Green Lake and Springs,	105
Te Rangipakaru Fountain,	107
A Sunset at Lake Rotomahana,	110
The Pink Terrace, or Tu Kapuarangi,	111
Ohinemutu to Orakei-korako and Lake Taupo by the Paeroa Range,	118
The Paeroa Plain,	121
The Hot Springs of Orakei-korako,	124
The Alum Cave,	133
Lake Taupo and Tapu-wae-haruru,	137
The Crow's Nest,	143
The Huka Waterfall,	144
The Bitter Lake,	146
Tokano,	149
Tongariro Volcanic Mountain,	151
White Island, Bay of Plenty,	156
'On the Influence of Atmospheric Changes on the Hot Springs and Geysers in the Rotorua District,' by Captain G. Mair,	158
Analysis of Water from a few of the Medicinal Springs in the Hot Lake District, by the Government Analyst,	162

INTRODUCTION.

A VISIT to the Auckland Lake district at the present time is quite a different affair from what it was a few years ago.

Since the first edition of this work was published, our railways, our coasting steamers, our mail coaches, and our inland hotels have become realities in New Zealand; and instead of having to 'do' the Lakes on foot or on horseback, with very great inconvenience, the travellers, whether ladies or gentlemen, can now sit down comfortably in the railway carriage, the steamer, or the stage coach, and away they go to this wonderland and sanatorium of the Pacific as on a pleasure excursion.

The extent of country comprised in this 'Hot Lake' district, measuring from north to south, or from the volcanic mountain Tongariro, near the middle of the North Island, to Whaka-ari or White Island, another active crater in the Bay of Plenty, is in length about one hundred and twenty miles, and the average width from ten to fifteen miles.

The distance eastward from Auckland to the Lakes

as the crow flies is about ninety miles, and the time at present required to perform the trip is as follows:—

From Auckland—two days, by rail and coach.

From Napier—three days, by steamer and coach.

From New Plymouth—four days, by steamer, rail, and coach.

From Wellington—six days, by steamer, rail, and coach.

Invalids from all parts of the world invariably come to Auckland as the nearest seaport to the Lakes, at which place they have daily opportunities of proceeding thence. Dr. Hochstetter, speaking of this Lake district, says:—'Over this whole distance, almost on the very line between the two active craters, it seethes and bubbles and steams from more than a thousand crevices and fissures that channel the lava beds of which the soil consists, a sure prognostic of the still smouldering fire in the depths below; while numerous fresh-water lakes, of which Taupo, twenty miles in diameter, is the largest, fill up the large depressions of the ground.

'This is the Lake district, so famous for its boiling springs, its steaming fumaroles, solfataras, and bubbling mud-basins, or, as the natives call them, Ngawhas (boiling springs) and Puias (fire springs), which until the last few years none but missionaries, Government officers, and some few daring tourists have ventured, by the narrow Maori paths through bush and swamp, to visit.'

Hochstetter goes on to say, in what has proved prophetic language: 'All those who have witnessed with their own eyes the wonders of nature displayed here, have been transported with amazement and delight; although only the natives have hitherto made practical use of these valuable medicinal springs, which are the grandest in the world, and they alone have sought relief in them for their various complaints and diseases. But when once, with the progressive colonization of New Zealand, these parts have become more accessible, then thousands dwelling in the various countries of the Southern Hemisphere, in Australasia, Tasmania, or New Zealand, will flock to where nature not only exhibits such remarkable phenomena in the loveliest of districts, with the best and most genial of climates, but has also created with these attractions such an extraordinary number of healing springs.'

In our local press, frequent notice is made of decided cures effected on the many helpless invalids who are, from almost all parts of the world, coming 'to be cured of their diseases,' the facility of transit, and the comfort provided, enabling them to accomplish the short journey with pleasure.

Referring our readers to the carefully compiled maps of this district for more particular information as to the position of the different lakes and springs, and the roads leading to them, we will now give a general outline of the wonders to be seen, and how best to see them.

Sketch of the Hot Lake District.

LAKE ROTORUA.— The first place to visit is Ohinemutu. This is a large native village, of great antiquity, on the south bank of Lake Rotorua. The coach roads from Auckland, Tauranga, and Napier all meet at this village; there is very comfortable accommodation at either of the hotels, and the village itself is full of hot springs, steam-holes, mud-pools, and natural bathing places, each one valuable for some particular complaint. The wonderful cures effected form the staple gossip of the place; but to our taste the best of all bathing places is the Rotorua lake itself, for we experienced it, at a good swimming distance from the shore, just sufficiently warm, from the numerous streams of boiling water that run into it, to be intensely luxurious, but exceedingly enervating.

SULPHUR POINT.—About a mile eastward from Ohinemutu, along the banks of the lake, is Puarenga, or what is called Sulphur Point. This is a most interesting place for invalids. One pool near the beach—a filthy-looking place, full of very hot water—is called 'Cure All;' the water is like soap-suds, and the smell is something to be remembered. Another pool, about thirty or forty yards from the beach, almost hid among the Manuka scrub, is called 'Pain Killer;' the water is quite clear, smells and tastes very strong, but the feel of this water in the hand is thick and clammy. Strangers are cautioned, by a notice attached to the frames of the bath, not to bathe in this spring without a friend standing by, as the water causes the bather to faint, it is so strong medicinally. The water in the baths here seem to be much stronger than at Ohinemutu; the ground in the vicinity is all incrusted with bright yellow sulphur crystals, and the peculiar smell is very palpable to the senses. Friend Anthony Trollope's 'putrid stench,' the name he applied to the smell, came to our recollection while threading the dangerous footpaths among the great mud-holes, sputtering up pancakes, basins of boiling sulphur water, and great steam-holes roaring as they send out clouds of steam all over the place, enveloping visitors in a London-fog-like douche bath.

Looking on these boiling caldrons, with their slimy green ooze, environed with such treacherous footing and abominable stench, one is reminded of Dante's grand picture of the rivers of fire and water in Hades :—

> 'We crossed the circle to the other bank,
>   Near to the fount that boils, and pours itself
>   Along a gully that runs out of it.
> The water was more sombre far than purple black;
>   And we, in company with the dusky waves,
>   Made entrance downwards by a path uncouth.
> A marsh it makes, which has the name of Styx,
>   This tristful brooklet, when it had descended
>   Down to the foot of the malign grey shores.

>   And I, who stood intent upon beholding,
>   Saw people mud-besprent in that lagoon,
>   All of them naked and with angry look.
>   They smote each other, not alone with hands,
>   But with the head, and with the breast and feet;
>   Tearing each other piecemeal with their teeth.
>   Said the good Master, Son, thou now beholdest
>   The souls of those whom anger overcame:
>   And likewise I would have thee know for certain,
>   Beneath the water people are who sigh,
>   And make this water bubble at the surface,
>   As the eye tells thee wheresoe'er it turns.
>   Fixed in the mire, they say, "We sullen were
>   In the sweet air, which by the sun is gladdened,
>   Bearing within ourselves the sluggish reek;
>   Now we are sullen in this sable mire."
>   This hymn do they keep gurgling in their throats,—
>   For with unbroken words they cannot say it.
>   Thus we went circling round the filthy few,—
>   A great arc 'twixt the dry bank and the swamp,—
>   With eyes turned unto those who gorge the mire.'
>   *Inferno*, vii. 100.

> 'From rock to rock they fall into this valley,
>   Acheron, Styx, and Phlegethon they form;
>   Then downward go along this narrow sluice,
> Unto that point where is no more descending,
>   They form Cocytus.'   *Inferno*, xiv. 115.

To make the picture more complete, Dante says that the Master

> 'From off His face He fanned that unctuous air.'

WHAKA-REWA-REWA.—About two miles from Ohinemutu, passing the place we have just described, is Whaka-rewa-rewa. It lies about half a mile inland from the lake, and is close on the banks of the river. The road from Ohinemutu is quite level all the way.

This place has wonders of a more extensive character than either of the two places we have named. The great clouds of steam that constantly hang over Whaka-rewa-rewa are visible from a great distance (from eight to ten miles). This will eventually be the great hospital of the Lake district, the bathing places are so numerous, extensive, and varied in their character. Here there are great mud-pools, twelve to fifteen feet across, throwing up hot, soft black mud from five to ten feet, in lumps about the size of a millstone; then basins of furiously boiling water, some of the water like thin soft-soap; others black, fetid, and semi-transparent, but with a smell doubly concentrated; and very close to this scent-bottle is a large pool of pure transparent water, slightly alkaline to taste, which, when cooled, is just drinkable, and strongly recommended for rheumatic patients. One little spring, just alongside a great boiling caldron, is exhibited by the guide with some considerable amount of mystery; it is about a foot deep, and has just a very tiny rill of pure cold water running from it into the great boiling pool on the other side of the footpath. We were informed that this insignificant spring has proved an effectual cure for long-standing diseases of the kidneys, etc., for many generations. We drank a cupful of the water; it was not disagreeable, but it was peculiar,—not to be described otherwise than water strongly medicated. It was here that Anthony Trollope saw water boiling with great *ferocity.* Poor Anthony got quite bewildered, and had to employ a native to lead him by the hand along the narrow footpaths. He says in his book that he 'became aware that a very *slight* spring, just one step forward, would not only destroy him, but it would destroy him with *terrible agony.*' And yet this novel writer, after leaving Whaka-rewa-rewa and all its terrors, writes : ' I have observed, all the world over, that the world's wonders, when *I have reached them,* have been less than ordinarily wonderful.'

TIKITERE SULPHUR SPRINGS.—The next place of interest to tourists is Lake Rotoiti and the Sulphur Springs at Tikitere. These are about six miles north from Ohinemutu. 'One case of recovery from rheumatism by means of these waters is worthy of quotation. A gentleman arrived here from Victoria thoroughly prostrated by this disease, in fact he had to be carried into the bath. After a week's systematic bathing he was completely recovered, and departed from Tikitere a strong, active *young* man, bearing with him the soubriquet of " Tikitere."'

Addison tells us of the curious properties of the Cave of Trophonius, that made people young again after spending one night in it. A week or a month at Tikitere will answer the same purpose.

ROTOMAHANA.—The next place of interest is Rotomahana, the wonder of wonders in this wonderful district. Travellers generally leave Ohinemutu in the afternoon to enable them to arrive before sundown at the Wairoa (Terere), a distance of nine miles, where there is good accommodation in the hotels. Here they remain for the night, about half a mile from Lake Tarawera; and as early as possible next morning the traveller will embark in a canoe (or whaleboat), large enough to accommodate in calm weather twenty, thirty, or even fifty people; and in three hours the natives, generally females, will drive the canoe across Tarawera to the Kaiwaka creek, which is about one mile distant from Rotomahana. They may spend one day or a week amongst the Terraces, and then return to the hotel of Ohinemutu.

ROTORUA TO LAKE TAUPO.—There are two roads leading from Ohinemutu to Taupo,—the coach road by Horo-horo, crossing the Waikato by the bridge near Niho-o-te-Kiore; the other road is only a footpath, and runs along almost in a line with the telegraph wires from Ohinemutu. Along the west side, and close under the igneous Paeroa range, which is full of all sorts of curious springs, boiling wells, and waterfalls, mud fountains, and dangerous footpaths, right on to the Waikato river at Orakei-korako. The natural wonders of this district eclipse all the others in number, variety, and magnitude,—the curious stalactite fringes, the enamelled floors, the immense number of boiling-water fountains, and the marvellous alum cave.

From this place the tourist will proceed to Tapu-wae-haruru, a rising village at the north end of Lake Taupo. There is here good, comfortable accommodation both for man and horse in the township; and a few days may be well spent in visiting the great natural curiosities of the district, such as the Tuahara Mountain, the Bitter Alum Lake, the Huka (Snow) Waterfall, and the Government bathing establishment.

TOKANO.—Across Lake Taupo by steamer, whale-boat, or canoe, or on horseback along the eastern shore, the traveller comes next to Tokano, the medicinal springs and baths of which are described by Dr. Hochstetter with great enthusiasm. Leaving Tokano, the traveller will rise in the world in proceeding along the banks of the upper Waikato and Poutu rivers. Passing Lake Rotoaira, he will ascend the great mountain Tongariro, up which the first tourist, the chief Ngatoroirangi and his slave Ngauruhoe, went with results most disastrous; he will learn how the mountain was originally set on fire.

These are a few of the natural wonders of New Zealand we intend to describe in this work.

FROM AUCKLAND BY THE WAIKATO TO THE HOT LAKES.

IN giving a description of the route to the Lakes by the Waikato, strangers will expect to get some short account of the savage war which had to be fought before we obtained possession of the fertile district through which this river flows. And in passing along by the railway the officials will call out names of stopping-places 'where battles were fought and fields were won,'—fields that will gladden the eye and the heart of the traveller, for they are nearly all turned into good, useful corn-fields,—Drury, Hunua, Tuakau, Pokeno, Mercer, Meremere, Rangiriri, etc., each one having a sad tale to tell of some incident in the great struggle.

We will make three outline extracts from our unpublished *Story of the Maori Wars*,—'General Cameron's First Fight in the Waikato—the Battle of Koheroa,' 'The Battle of Rangiriri,' and, in connection with the other route, *viâ* Tauranga, 'The Massacre at the Gate Pa,'—a little sensational reading, which will help to while away the time.

Only a few years ago, the whole of the Waikato country (upper and lower) was in the undisputed possession of the Maori. This district in length from north to south measures over one hundred and twenty miles, and the average breadth from east to west is about one hundred miles. The quality of the land is quite equal to that of any district in the colony. This fact will be apparent to the traveller as he passes through it on his way to the Lake country.

The Waitara (Taranaki) disputed land was abandoned by the Government on May 13, 1863, on its being found that the Waikato tribes were supplying the fighting-men for Taranaki. When remonstrated with, they excused themselves by saying that the Ngati-manio-poto, the great fighting tribe of Waikato, by right of conquest made many years ago, now laid claim to the whole of the Taranaki district.

General Sir Duncan Cameron, with the concurrence of the New Zealand Government, immediately began preparations for a great and final struggle,—the invasion and subjugation of Waikato. This would have been more decided and complete if sufficient time to perfect the preliminaries had been obtained; but these were cut short abruptly by the receipt of an authentic report which reached the Colonial Government on July 8 the same year, that on a certain day (which was named) Auckland city was to be set on fire in several places (the places were even named), and during the confusion it was to be attacked, the inhabitants massacred or driven into the sea, and the place itself plundered. In fact, at this time this grand programme could have easily been carried out, for a large number of Maoris and half-castes were living in the city.

An immediate forward movement of the troops was at once made, and on the 12th July, at daybreak, Colonel Wyatt took possession of Tuakau, a native village within sight of the Waikato river. On the 10th, General Cameron had advanced to what is called the Bend of the River, now the Queen's Redoubt (this place overlooks the Waikato, and is a most commanding position). On the 12th, the same day that Tuakau was taken, he crossed the Maunga-tawhiri river, and encamped on the Koheroa ranges. This was an invasion of the enemy's country, as this river, like the Gate Pa, formed the boundary line between the native and European lands.

On the 16th, Colonel Murray surprised Keri-keri, and took thirteen Maori prisoners (whom we had to feed and guard for a few months, until, getting tired of this, they were allowed quietly to escape). On the 17th, the Maoris and the British soldiers for the first time met face to face at a fair stand-up fight.

These preliminary notes will enable strangers to understand the following account of—

GENERAL CAMERON'S FIRST ENGAGEMENT IN
WAIKATO—THE BATTLE OF KOHEROA.

No sooner had the general and his men crossed the sacred boundary, than it became evident that the enemy had taken up a position with the view of opposing the advance of the British troops towards the Whanga-marino river, and where the path led by narrow ridges they had strengthened their position by digging rifle-pits. As the troops advanced, the Maoris gradually retired from their first line, which had only been partially formed; and this allowed our forces to gain on them by degrees.

On coming within range, Captain Strange, 14th Regiment, with his company, ran rapidly forward and occupied a ridge on the right of the enemy's retreat. The latter halted immediately under cover of the crest, and opened a sharp fire across the gully on our skirmishers, who rapidly replied. Our main body followed the line of the enemy's retreat, and on reaching a small knoll, within one hundred yards of their second line, were received with a rattling volley, which, by its suddenness and severity, for a moment checked the young battalion. It was their first time under fire.

Seeing how matters stood at this critical moment, General Cameron rushed forward at least twenty yards in advance of his men, and waving his cap in the air, cheered them on, calling on his troops to turn the Maoris at the point of the bayonet. The order of the gallant veteran was instantly obeyed, and with a true British cheer the enemy's position was rushed. Many of the Maoris stood in the trenches, and fought well and manfully, but that inimitable weapon, the bayonet, in the hands of a British soldier soons settles the business.

As the soldiers advanced, the Maoris, running in dismay to the nearest cover, sprang into a ravine to their right. At this juncture, the soldiers having rapidly formed round the semicircle of high ground which embraced the ravine, poured in a murderous converging fire on the enemy; and as they fled through the bottom of the ravine, many of them fell, while others again, keeping on the high ground, retreated to a farther ridge, where they again opened fire on our advanced forces.

The troops, led in person by Cameron, again drove them from their vantage ground, and at length, broken and disheartened, they fled before the soldiers to the Maramarua river, which some crossed in canoes, others by swimming.

The forces were then withdrawn to the camp at Koheroa. This engagement began at eleven A.M., and ended at one P.M. The enemy numbered 400; British, 21 officers and 532 men.

This was the first great battle in the Waikato campaign. The engagement at the Gate Pa was the last. The one was as brilliant as the other was disastrous. All things considered, this first engagement seems to have convinced the leading Maori chiefs that their case was hopeless; for in the very next engagement, that at Rangiriri, the King Potatau and his chief adviser, Tarapipipi, although both were actively engaged at the beginning of the fight, contrived to make their escape some considerable time before the white flag was hoisted and the garrison made prisoners.

After this affair, the general advanced rapidly into the Waikato country by the river highway. The first steamer on the Waikato river, the *Avon*, reached the Bluff stockade on the 27th July, and on August 14th the troops advanced from Koheroa to Whangamarino. A second steamer, the *Pioneer*, arrived on October 3. The next fortified position on the Waikato was Mere-mere.

THE BATTLE OF RANGIRIRI.

On the 29th October, Cameron went up the Waikato in the *Pioneer* steamer. On passing Mere-mere, the enemy turned out in considerable numbers, and rapidly, but in silence, occupied the tiers of rifle-pits. In silence also the steamer passed on, without those on board taking the slightest notice of the Maori and his rifle-pits, and went about twelve miles farther up the river to Rangiriri. At this place the Waikato is about two hundred and fifty yards wide, with soundings from nine to eighteen feet, and a current running at the rate of three miles.

The Rangiriri stronghold was situated very low, just a common embankment thrown up with a trench cut in front of it. After a careful examination of this place, the steamer was turned round, and went down the river again. On passing Mere-mere, one of the Maori big guns from the entrenchment was fired at the steamer. The charge had been a seven pound grocer's weight, and it lodged in a cask of beef which happened to be on the deck of the steamer.

On November 20, General Cameron went up the river again, determined this time to reduce Rangiriri. At three P.M. they sighted the enemy's entrenchments, about six hundred yards from the works. The extent of their front line was five hundred yards, and at the highest point (the centre) was strengthened by a formidable redoubt, having a ditch twelve feet wide, and parapet eighteen feet to the top from the bottom of the ditch. One of the steamers, the *Pioneer*, got stuck in the river, with 300 men of the 40th Regiment on board.

The order of attack was 200 men of the 65th under Wyatt, 72 men of the 65th under Lieutenant Toker, and a party with scaling-ladders and planks under Captain Brooks. At half-past three in the afternoon the enemy opened fire along the whole line, without damage. The order to land the troops from the steamer *Pioneer* was made in vain. So General Cameron, at five P.M., determined to wait no longer, gave the order for the assault, and the troops dashed at it. At about fifty yards the skirmishers were ordered to halt and cover the ladder party while planting their ladders. While the 65th were scaling on the left, the 12th and 14th were to keep down the fire in the centre. The broken ground and the heavy firing tried them severely during the advance. Colonel Austen and Captain Phelps of the 14th, and many others, were wounded almost immediately on becoming exposed. The enemy's fire was sharp, quick, and heavy; but nothing could check the assault. The ladders once planted, the 65th immediately forced their way into the enemy's works.

The first line carried, the natives fell back on the centre redoubt and adjacent works; and just at this time the 40th landed from the steamer, and were sent at once to the ridge in the rear, which they at once carried, and the natives fled for the swamp of Waikari, in attempting to cross which several were drowned, or perished under the fire of our rifles. A part of the 40th held the hill, the remainder joining the main body under Cameron.

The troops now closed on the centre redoubt, where the Maori fought with desperation, and held his ground against every attempt to dislodge him. Two distinct assaults were made on this work. The first by the Royal Artillery, with revolvers, led by

Captain Mercer, who received a severe wound through the jaw and tongue; and every man who attempted to pass the place where he was lying was either killed or severely wounded, all except one man, Lieutenant Pickard. Mercer and the other wounded men could not be removed for some time.

A second assault was made by 90 seamen, also armed with revolvers, led by Captain Mayne of the *Eclipse*. They went against the front of the works, and were received with a deadly volley, which prevented them from effecting an entrance. It was now dark. General Cameron gave strict orders that the troops should remain in their respective places, in which position they almost surrounded the enemy; and thus they remained all night. Long and anxious consultations were held that night; and Colonel Mould had just commenced, next morning at daybreak, with pick and shovel to dig the parapet down, when a white flag was hoisted by the Maoris, and they surrendered unconditionally.

Rangiriri surrendered on November 21, and on December 8, 1863, Ngaruawahia was occupied and the Waikato virtually won.

The Whanga-marino river is now crossed by a railway, and the Waikato at Ngaruawahia, the ancient Maori capital, is also bridged by a magnificent structure raised on immense iron tubes, and elevated far above the highest high-water mark. The railway, after crossing the bridge, runs through the village, passing close by the tomb of Potatau, the first Maori king of New Zealand.

The village, now quite English in its character and appearance, lies on a fine rich flat running back from the rivers; for it is pleasantly situated between the two, the Waikato and the Waipa. In the middle of the village is carefully preserved the tomb of the poor old king; it is fenced in and planted with trees and shrubs, the outer circle just inside the fence being a row of British oak trees.

The town of HAMILTON is divided by the river into east and west, but connected by a substantial bridge, and is now a pretty considerable town.

CAMBRIDGE is at present on the borders of the King country. The traveller will observe that the higher he gets up in the Waikato country the land improves in quality; and from the neighbourhood of Cambridge the extent of easily available pasture and agricultural lands seems unlimited.

AUCKLAND TO TAURANGA—SEA VOYAGE ABOUT TWENTY HOURS.

As the coasting steamer for Tauranga harbour takes cargo as well as passengers for the intermediate settlements, a good opportunity will thus be afforded for seeing these places. We will therefore give a description of the voyage to Tauranga. The steamers generally leave Auckland harbour about five o'clock in the evening, so as to reach Tauranga next afternoon before dusk.

As the captain gives the order, 'Easy ahead' and 'Cast off,' the steamer, laden with passengers and cargo, glides gently down Auckland harbour. Away on the clear, sparkling Waitemata, and we soon find a goodly proportion of our fellow-passengers are, like ourselves, also off to the Lake district.

Round the north head of Auckland harbour, past Rangitoto Island, round Cape Colville, and about daylight next morning we are threading our way among the Mercury Islands.

Passing the odd-looking, isolated obelisk, one hundred and twenty-five feet high, called Man Rock, on the one side, and the equally curious Needle Rock on the other, we come in sight of Mercury Bay.

In Mercury Bay we are on classic ground, for it was here that *our* great navigator, Captain Cook, after searching throughout the length and breadth of the South Pacific for the mythical southern continent, came to anchor, and going on shore with his rude astronomical instruments, observed and noted the transit of the planet Mercury over the sun's disc, on the 9th November 1769.

It is curious to recall the remarks made by Cook about what he saw, and so carefully examined, during his visit to Mercury Bay. Let us see what great changes the winds and the weather have made since the *Endeavour* lay at anchor near Shakespeare's Cliff, while Captain Cook 'displayed the English colours, and took formal possession of New Zealand in the name of His Britannic Majesty King George the Third.'

Cook says: 'There are several islands both to the southward and northward of the bay, and a small tower or rock (this is Motu Korure) in the middle of the entrance.' 'The best anchorage is in a sandy bay (Cook's Bay) which lies just within the south head, bringing the high tower or rock (Motu-roa), which lies without the head, in one with the head, or just shut in behind it.' These are, to all appearance, unchanged.

The north entrance point of Mercury Bay is a perforated rock about half a mile from the mainland. Cook's description of this rock will show what a wonderful change it has undergone in the course of a century, and how the natives have decreased and disappeared from this once populous place.

Cook says: 'Of two fortified villages, we landed near the smallest, the situation of which was the most beautifully romantic that can be imagined. It was built upon a small rock detached from the mainland, and surrounded at high water. The whole body of this rock was perforated by a hollow or arch, which possessed much the largest part of it. The top of this arch was more than sixty feet perpendicular above the sea, which at high-water flowed through the bottom of it. The whole summit of the rock above the arch was fenced round after their manner, but the area was not large enough to contain more than five or six houses; it was accessible only by one very narrow and steep path, by which the inhabitants at our approach came down and invited us into the place.'

The native name of this perforated rock, like all other native names of places, is descriptive of its character—Mohungarape, the literal meaning being 'mealy,' or 'crumbling like a boiled potato.' And how truly appropriate is this name to that weather-beaten rock at the north entrance point of Mercury Bay, the place that contained 'five or six houses' being now so narrow that one could sit astride as on horseback!

Cook says: 'The safest place for a ship that wants to stay here any time is in Mangrove river. To sail into this river, the south shore must be kept on board all the way.' The eastern side, on entering the river, is a barren rocky bluff, and just inside, in a small bay, are the buildings of the unfortunate Auckland Sawmill Company.

On the west side or right hand the prospect is more pleasing,—a fine hard sand-beach, backed by well-fenced paddocks and clumps of native and European trees. There are a few neat houses surrounded with gardens, and a large hotel now occupies the site of the Huki-huki Pa (a place for roasting fish). The Mangrove river winds through the valley, and away in the distance is seen heavily timbered land.

At the head of Mercury Bay, partly embedded

in beach sand, are still to be seen the timber ribs of Her Majesty's ship *Buffalo*, wrecked there so long ago as March (Terry says August) 1840. A few lazy sea-gulls sit winking on top of them, waiting for some enthusiastic *genre* painter, or a second Dr. Syntax in search of the picturesque, to come and take a sketch of Pai Marire.

A few hours' sailing and we are off Tairua river, the navigation of which is difficult and dangerous. The steamer passes inside the Shoe and Slipper Islands, so wonderfully shaped, the one like a well-made shoe, the other like an ill-made slipper with the heel down. As the steamer glides along quite close to the shore, several old native settlements will be observed, nearly all deserted, with the exception of a few old men and one or two shrivelled-up female atomies, but not the ghost of a child, either boy or girl, is to be seen. This is the most marked change in New Zealand.

When Captain Cook passed along, over one hundred years ago, crowds of natives greeted him everywhere. On every hill-top from which his vessel was visible, at every promontory which it passed, crowds of natives of all ages sat gazing at the wonderful machine, full of human beings so like themselves in shape though not in colour. It is quite probable that the population of New Zealand at that time was not less than 150,000 or 200,000; and what is it now? certainly not much more than 30,000.

TAURANGA HARBOUR.

Having passed Whangamata and Kati-Kati early in the afternoon, we made Tauranga harbour. As already stated in our *Traveller's Guide through New Zealand*, for picturesque beauty this harbour is unsurpassed in the colony. It is easy of access, and perfectly safe inside. The country on both sides of the harbour is exceedingly fertile, with a deep, rich alluvial soil.

From Te Papa or Tauranga village the peninsula extends away inland, on which stood the celebrated Gate Pa and Te Ranga Pa, or rather where they once stood, with their mysterious fencing and rifle-pits defying thousands of our best British troops, led on by our most experienced commanders. But it is all over now, for the one Pa has been levelled, ploughed, and turned into a corn-field, while the site of the other is occupied by a farm-steading and dwelling-house. And now, where the brown fern grew, and the brown Maori considered himself supreme, the land has been laid off into paddocks, fenced and put under cultivation; and the traveller, on looking round from the top of the hill where the Gate Pa stood, will see in every direction dwelling-houses, gardens, orchards, schoolhouses, and churches, giving evidence of the energy and enterprise of the new race of settlers.

The township is rapidly taking shape and form. The main street is broad, level, and clean; runs parallel with the river; and along the whole length, less than a mile, there is now a goodly show of buildings, hotels, stores, shops, banks, offices, and workshops; and running well into deep water there is a substantial wharf.

Tauranga to Ohinemutu by Stage Coach.

The mail coach leaves Tauranga early in the morning, picking up passengers at the hotels in passing through the town; and, after settling quietly down into your seats, *almost* prepared for the long wearisome journey over one of the roughest roads in the colony, when about three miles from town the coach is stopped at the top of a gentle rise, the driver jumps down, and, quietly proceeding to light his pipe, he looks up to the passengers, says, 'Ladies and gentlemen, would any of you like to see the Gate Pa?' At these magic words we look round, but no Pa is visible,—nothing but fields of wheat, oats, potatoes, and settlers' houses; however, we jump down, and the driver points to a ditch and parapet overgrown with Tupakihi and furze. We look at it and then at the driver with disgust, and express anything but satisfaction at the hoax. So we get up into our seats again, and the driver, after fixing himself on the box, begins to tell us the true and faithful account of what happened at the storming and capture of the Gate Pa.

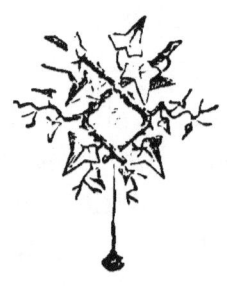

THE STORY OF THE MASSACRE AT THE GATE PA.

WHERE the Gate Pa stood the land is narrowed by a deep swamp on both sides; from these the ground rises gradually until the summit is attained. It was here that the main Pa stood, stretching right across the road which led into the interior of the country; this is the line of road now traversed by the coach.

From the general level of the land, the elevation is slight,—only about fifty feet,—and the ascent in front is not difficult. On both flanks of the redoubt or Pa the approaches were even, the land inclining towards the swamp rather above than below the level of the flanking trenches.

The ditch or trench, about three feet deep, which extended from swamp to swamp, that had been cut as a boundary line between Church Mission and the native land, had been cleared out and deepened to the swamps, the door or gate was near the centre, and no one could get into the interior of the country but through this gate, and hence the name.

On the left flank the natives had constructed three lines of traversed rifle-pits, and surrounded them with a Ti-tree wove fence, which enclosed a space of eight yards wide by thirty yards in length. These works and a ditch seventy yards long protected their left front.

The main Pa, connected in the way described with the flanking redoubt, was constructed in precisely the same way. It was about eighty yards in length by thirty in breadth in the centre, and twenty at the flanking corners. There were three tiers of zig-zag rifle-pits communicating with each other, and these were roofed with Ti-tree wattles and thatched with fern, and in some cases they were also covered with earth.

The stockading of the Pa was first a post and then rail fence, and four feet at the rear of this a Ti-tree wove fence. Behind these were the first lines of roofed rifle-pits, communicating with those in the rear. The above description of the works may not be very intelligible, but it will show, from the result of the struggle, that considerable skill had been displayed in their arrangement and construction.

On the morning of April 28, 1864, General Sir Duncan Cameron, with three regiments of infantry, sappers and miners, and marines, numbering over 4000 men, took up position in front of this mysterious fortification. Cameron had with him all the appliances of modern warfare, but he was in complete ignorance of the character or nature of the works to be attacked, and of the natives who crouched behind them watching unseen his grand military movements.

Our troops had one Armstrong gun of 110 pounds,

and during the day fired, it is said, 100 shells at the paltry fence we have just described. The settlers in the neighbourhood have turned up a number of these, as well as smaller ones, in ploughing their fields, and all with the dangerous charges still in them. In a paddock a long way beyond the field of battle we examined among others one of 110 pounds; the outer cast-iron shell was whole, and the inner casing of zinc with the exploding cap waiting for some accidental knock with a hammer, a harrow, or ploughshare to do some terrible mischief. It is only a few years ago that one of a road-making party turned up a 24-pound shell, and an inquisitive labourer thought to open it by turning round the cap with a hammer and chisel. It went off, but it took the labourer's head with it, knocked off one of his feet, and smashed the legs of a looker-on. Now, who is to blame for leaving these dangerous playthings lying about?

There were about 500 natives in the Pa, but they contrived to deceive our forces by raising several red fighting flags on long poles, and distributing them in the open ground away behind the stockade. It was against these imaginary fortifications that the greater number of the shells were directed, and of course they sunk into the soft earth without effect. The whole force of the natives was thus enabled to man the trenches in front, and make General Cameron believe that the enemy were at least 2000 strong.

The cannonading continued until nearly nightfall, when the 68th Regiment, having got to the rear of the Pa by skirting along the edge of the swamp, and a breach having been made in the centre in front, a storming party was told off to rush the place.

Captain Hamilton with the 43rd Regiment were the first at the breach; he sprang upon the parapets, and shouting, 'Follow me, men,' dashed into the fight; but that moment a bullet pierced his brain, and he fell dead, and many of his officers shared the same fate. There was a momentary lull outside.

Thrice the column of the 68th attempted to charge up to the proper right of the enemy's position to take it in reverse, and thrice they reeled and fell back. The official despatch says, 'This was not produced by any resistance on the part of the natives, but solely from the cross fire of our own men.' This statement explains a strange rumour we have heard more than once, that the two regiments, the 43rd and the 68th, had been at deadly enmity during the campaign in the Waikato, and that they had to be separated several times by their commanding officers. They were actually firing at one another instead of at the enemy!

To assist the reader to understand how this cross fire became so disastrous, and was probably the principal cause of the terrible repulse or massacre of the British troops, observe that the firing from the great guns, etc., during the day had been from the camp on a gentle rise about half a mile or so back, but almost on a level with the Pa. This was on the right. The Naval Brigade and the 43rd, comprising the storming party, having formed under the brow of the hill on the left of the enemy's works, were led in gallant style at the double *across the flanking fire* in the pits on the left, and into the breach.

The stormers, led by Captain Hamilton, as we have already said, entered the breach with a cheer, which was answered by their comrades over the field, and with another loud hurrah, the 68th, having crossed the swamp, charged up the hill in rear. Fire was opened by the 68th and the detachment of the flying column under Ensign Cartwright on the opposite hill, but the 68th, owing to the heavy cross fire to which they were exposed at short range *from both friend and foe*, were compelled to retire. The 43rd and Naval Brigade lost nearly all their officers, for they were either killed or wounded.

The explanation of the whole matter is, that when the 43rd charged in front, the natives, finding it much too hot for them, attempted to escape by the rear, and were nearly all out of the Pa when the 68th and flying column came suddenly and quite unexpectedly upon them from behind. The natives were thus compelled to return to the Pa and fight to the death. And as these natives came pouring in from the rear, *some one* cried out that the natives were on them

5000 strong. Although at this time the soldiers had possession of the place, a panic seized them, and they fled. 'Such, ladies and gentlemen,' said our driver, 'is the true story of the Massacre at the Gate Pa.'

The road from Tauranga to Rotorua, about forty miles in length, may be called a good road for strong, healthy travellers, but the alarming jolts and shakings are scarcely suited for sickly people. A lady has published her experience; she says, 'Eighteen miles of bush we traversed over logs and stones—bump! crash! It was a good thing I had neither false hair nor false teeth, for these must inevitably have been scattered to the winds;' and Mr. T. L. Travers says, 'The scenery would be more enjoyable if the ruts were less than eighteen inches deep, and the roots and logs removed.'

About five or six miles from Tauranga the driver will point out where Te Ranga Pa stood, now occupied, as we have said, by a farm-steading. It was captured June 21, 1864. He will probably tell another long story about its capture, and how many Maoris were killed, for he will refuse to tell you how many were killed at the Gate Pa. And so, passing along the green banks of the Waimapu river, a long steep rise will bring you to the edge of the Oropi bush about thirteen miles from Tauranga.

THE OROPI EIGHTEEN-MILE BUSH.

THE New Zealand bush, as Judge Avney said of the Hot Lake country, can be seen and admired, but to describe either the one or the other is a more difficult task. A photographer cannot do it, neither can a painter with his brush and box of colours, although the human eye can take it in at a glance. Near at hand bush scenery is graceful and pretty; while farther back, away in the gloomy distance, where the sunlight never penetrates, or away up aloft with the giants of the forest, it is awfully grand.

At your feet a dense growth of closely-matted delicate mosses and ferns, some of them very minute, with such graceful lace-like leaves, and bending over all this the tree-fern forms a canopy to the fairy scene. A little way off lies the dead trunk of a giant Pukatea rotting and covered with fungus, while the native convolvulus with its pretty blush-pink bells twine round it; and far away as the eye can distinguish among the dense undergrowth, are seen the bare shining stems of the big trees and the great tree-ferns luxuriating in their native element.

About half-way through the forest the grandest sight is Mangarewa gorge; it is some hundreds of feet in depth, and on the opposite side from the road the precipitous cliffs tower up rock-bound and tree-clothed in bold and majestic grandeur. Glinting rays of sunlight light up the whiteness of the rocky crags, and give relief to the sombreness of the dull green of the huge trees projecting almost at right angles from amongst them. Below, all is dark, damp, and dismal, and the limpid waters of the stream throw themselves noisily over the rounded boulders shrouded beneath a weird canopy of overhanging pines. No gleam of sunlight reaches these depths, and as the bridge is crossed the heavy darkness strikes cold and penetrating. Very charming, in spite of the wild gloom, is the glimpse of the river-bed winding away apparently into a dismal cavern of gnarled and ghostly-looking trees. It is a spot that solitude has marked for its own.

The road through the Oropi bush is said to be in length eighteen miles, but it is more like twenty-eight; and although the Kauri is absent, the Pukatea is as noble-looking a tree; it is, in fact, a magnificent pine. As it is here surrounded with the dense undergrowth in a sort of twilight, looking up at its height the eye gets bewildered. Although the foliage is neither pretty nor graceful, yet the tree is a giant in a forest of giants.

The beautiful light green fronds of the tree-fern (Punga), growing in the close, moist atmosphere, rise above the undergrowth, and almost encircle the giant stem to the height of twenty or thirty feet. Above this come the creepers, the Rata and the mistletoe rising up into daylight above the surrounding forest. These brilliant parasites hang from the uppermost boughs of the loftiest trees, or, winding from stem to stem with fantastic curves, interlace distant trees in the very extravagance of their luxuriant beauty.

The lofty Totara, and the Rimu with its delicate and gently weeping foliage, and the shade-loving fern-tree, the most graceful of all, add variety to the charming scene. Wild flowers are few and rare, but the ferns are more numerous and varied than in any other forest we have seen. But it is the absence of all living things which renders the silence and solitude of the New Zealand forest so oppressive at times. Occasionally a pair of Kaka parrots may be seen wheeling high above the tree-tops, with harsh, discordant cries, or the melancholy note of the great New Zealand pigeon comes booming through the woods; and on rare occasions the soul-stirring, deep-toned bugle note of the Tui (Captain Cook's parson bird) may be heard singing love-songs to his mate. But, except at early morning, the traveller may often wander for hours—we might say days—together through the gloom of these woods, where the sun's rays can scarcely penetrate, and the breeze passing over the tree-tops through the uppermost whispering boughs may be seen and heard, but cannot be felt.

TAURANGA TO MAKETU, 17 MILES; THENCE TO TAHEKE, ON LAKE ROTOITI, 21 MILES; THENCE TO OHINEMUTU, 12 MILES.

THIS is the first path traversed by Europeans to the Lake district; it is not now used by tourists, but as there are several interesting spots on the way, we give a short sketch of it. From Tauranga across the river to Matapihi settlement, then along the beach road to Kaituna creek, which is crossed at present in a punt, Maketu is a pretty settlement on the banks of the Kaituna river, which takes its rise at Rotoiti. The entrance to Maketu is called Ngatoroi, after a celebrated ancestor, Ngatoroirangi (who will be noticed farther on); he came over in the Arawa canoe. The stones to which this canoe was fastened are still to be seen at the heads, and are called Tu-te-rangi-harinu and Toka-parore; and on the beach a clump of trees still growing, called Angi-angi, which formed the skids of the canoe, having been left here they took root and grew to trees. After a ride of 21 miles over a very good road (made expressly for H.R.H. Prince Alfred by the Maoris), you arrive at the Taheke settlement on the banks of the Rotoiti lake, which is most beautifully situated, the view of the lake and its surroundings being truly magnificent. Here you will see an enormous carved house called Rangitihi, the walls of which are formed of carved figures of the ancestors of the Rotoiti people, and the skilled men who did the carvings have given each figure its full proportions. The refined taste of Anthony Trollope found them 'grotesque and indecent.'

About a quarter of a mile from the Taheke settlement the Ruahine springs are worth seeing. The following description of them is by a travelling auctioneer from the south (Christchurch):—' Gradually approaching the boiling wonders, we perceive the steam curling above us, and our nasal organs at once proclaim them to be strong in the odour of brimstone—so strong, that some of our party, thinking of

> ' " The undiscovered country, from whose bourne
> No traveller returns,"

refuse to proceed farther; however, a few yards of cautious travelling and our curiosity is more than satisfied, while our wonder is increased. We have now actually seen and can believe that such things as boiling springs do exist,—not merely hot water bubbling up on a level with the surface, but large pools of *galloping* boiling hot water.'

Leaving Taheke, a few minutes' ride will bring you to the bridge over which you cross the Kaituna river, which here presents a series of rapids both above and below. Following the road, you soon reach a native settlement called Mourea, where there is another bridge across the stream, which connects at this place the Rotorua lake with Rotoiti. These two lakes are only separated by a neck of low land, through which, in a stream, the Rotorua discharges its waters into the neighbouring lake. This stream is deep enough to allow a large whale boat to pass through from one lake to the other. Three miles farther on the road brings you to Te Ngae.

After reaching the summit of the hills beyond the old Pa, with its 'rude specimens of native carving,' a beautiful view is obtained on either side. To look back to see the beauties of Rotoiti, the bold headlands running far into its waters, its many islands, and the general outline of the surrounding country, form from this spot a lovely picture. To look ahead is to view the vast expanse of Rotorua, with its romantic island, its white cliffs, and the numerous columns of steam arising from the boiling springs of Ohinemutu and Whaka-rewa-rewa.

TIKITERE SULPHUR SPRINGS.

After enjoying the view, the traveller will turn to the east by a path leading about three miles along the south shore of Rotoiti to the sulphur springs of

Tikitere; these form a whole valley of bubbling mud-pools and boiling springs. In the middle of the valley is a water-basin 50 or 60 feet in diameter, called Huritini, which is constantly boiling, seething, and bubbling, the muddy water sometimes rising to a height of 15 feet. The whole valley is one mass of solfataras, only to be compared to hideous carbuncles on the surface of the body. They are, however, real sulphur wells, more or less deep, surrounded with yellowish white crust, and diffusing an offensive odour. The pumice sand is cemented with silicious deposits. The sulphur crusts and black mud form a very suspicious soil, which can only be stepped upon with the greatest caution; and the atmosphere, impregnated with sulphuretted hydrogen and sulphurous acid, which is ever whirling up in dense clouds of steam from the dismal-looking openings, makes a lengthened visit exceedingly uncomfortable. A recent traveller says:—'The Tikitere springs are situated about three and a half miles up the picturesque valley of Te Ngae. In visiting these, it behoves every one to be most cautious, as the ground all round is most treacherous. Any person finding the crust breaking under him, would do well to fall flat down as lightly as possible, and then crawl out until he arrives on *terra firma*—the firmer the better. In close proximity to the first set of springs are a lot of native huts. These are erected for the use of natives frequenting the springs for the purpose of bathing.

The curative properties of these waters are most effectual in cases of rheumatism or skin diseases. The springs first arrived at are active, boiling sulphur water of a brownish yellow colour; the banks are one mass of sulphur, some of the sulphur being quite pure and very fine. The next group are powerful boiling mud-springs, strongly impregnated with sulphur, and constantly active. We then followed along the bank of the creek, passing a small waterfall, until we arrived at a large hot sulphur lake ; its banks abound in small caves lined with crystal sulphur, the roofs being covered with stalactites of the same material. Beautifully formed specimens of crystal sulphur can be obtained from this lake.'

Returning to the main road, from which we diverged to see these curiosities, the traveller arrives at the old mission station of Te Ngae, where our respected namesake, Rev. Thomas Chapman, resided over forty years. Eight miles along a good level road, and the traveller arrives at Ohinemutu.

Lake Rotorua.

The circular form of this lake, the little island in the middle, the white steam-clouds ascending along the shores, all these might induce the observer to take Rotorua to have formerly been a volcanic crater, while in reality this, like all the other lakes of the district, has probably been produced by the sinking of parts of the ground upon the volcanic table-land. The depth of the lake is perhaps at no place more than five fathoms. It has numerous shallow sand-banks; and the shores, except on the north side, are sandy and flat. It is 790 feet above sea-level; while Tikitapa is 1198, Rotokakaha is 1127, Tarawera is 816, and Rotomahana is 834. The highest hill, that on the south-west side, the forest-covered Ngongotaha, is 1331 feet above sea-level, from the top of which can be obtained a most extensive view in all directions, reaching to the shores of the Bay of Plenty and Whaka-ari or White Island, to Lake Taupo, with Ruapehu and Tongariro, and westward to the Upper Thames, Piako, and Waikato. The largest river emptying into the lake, the Whaka-rewa-rewa, is on the south-east side. On the north-east the Ohua creek forms the outlet of this lake to Rotoiti, thus, as already noticed, connecting the two lakes by a narrow strait.

The centre of the hot springs is Ruapeka bay or Ohinemutu village; and the principal boiling spring, that on the south side, throws out a quantity of hot water which makes the whole bay quite warm, and forms the favourite bathing-places for both natives and Europeans, as by approaching the fountains more or less closely any degree of temperature may be selected.

In going between the numberless pools of boiling,

sputtering mud and water, the greatest care has to be taken. Dr. Hochstetter says, 'Whoever has once involuntarily bathed his feet in steaming water or boiling mud will certainly remember it all his life.'

OHINEMUTU.

Ohinemutu is a large native settlement belonging to the Arawas, and is a place of note in the history of this tribe. There is a native Runanga or meeting-house here, which is perhaps the best specimen of Maori architecture and carving in the colony. It is three times the breadth of the Maori house at Wellington, which was taken from Turanganui and attached to the Colonial Museum. It is named Tamati Kapua, after a great ancestor of the people. Domett gives a good idea of the carving on the panels:—

> 'Piled up on pillars, squat monsters rise
> Perched on each other's shoulders to the roof;
> The tribes' great council chamber this should be,
> Their Whare-kura, Hall of Sacred Red,
> For worship—justice . . .'—C. 8, p. 144.

Lake Rotorua is about eighteen miles long, and, seen from Ohinemutu, has the appearance of an inland sea.

One remarkable character to whom we were introduced was Hori Haupapa, an aged chief. Hochstetter met him some thirty years ago, and wrote of him as 'a giant in size and a Hercules for strength,'—a description which his still massive limbs would justify, although age has bent his body; he stood then six feet six. Many stories are told of his great strength. On one occasion he pitched an adversary clear over the high palisading of a Pa. He was defeated by the great cannibal Hongi, when he stormed and took the island of Mokoia. Hori escaped by swimming to the mainland. The old man does not speak much, but they said that he remembered Captain Cook; this would make him 115 years old. During our visit to Ohinemutu, a midnight Haka was held, the leading dancer being Hori Haupapa's daughter,—one of the most graceful, lithe dancers, and certainly the best and most agreeable-looking Maori girl, we have seen. This settlement has always been famed throughout New Zealand for the beauty of the women and the gigantic height and strength of the men, from the days of Hinemoa down to the present time. We observed that the younger girls have a complexion like the Spanish gipsy, just fair enough to let the warm colour show through the clear olive skin, large dark lustrous eyes, with great and ever-changing expression, rosy lips, beautiful snow-white regular teeth, and small, well-shaped hands and feet.

The village of Ohinemutu is built on a thin crust of rock and soil roofing over one vast boiler. Hot springs hiss and seethe in every direction, some spouting upwards and boiling with the greatest fury, others merely at an agreeable warmth. From every crack and crevice spurt forth jets of steam or hot air, and the open bay in the lake itself is studded far and near with boiling springs and bubbling steam jets. So thin is the crust on which these people have built their little homes and lived for generations, that in most places, after merely thrusting a walking-stick into the ground beneath one's feet, steam instantly follows its withdrawal. Food is boiled by being hung in a flax basket in one of the countless pools. Steaming and baking are performed by simply scraping a shallow hole in the earth wherein to place the pot, and covering it up again to keep the steam in, or by burying the food between layers of fern and earth in one of the hot-air passages.

A remarkable open-air service of the Church of England in Maori was held here on Sunday, December 18, 1870. 'It was a calm and beautiful day, and the scene was highly picturesque and suggestive,—Prince Alfred, the Duke of Edinburgh, the Governor of New Zealand, a few Europeans, and a large congregation of natives repeating the responses and joining in the hymns of the Church in their own sonorous language, amid the finest prospects of lake and mountain, and near some of the most wonderful natural phenomena in the world,—in the heart of the native districts and of the country most renowned in Maori song and legend, and where, in the memory of men still living, cannibal feasts were often held.'

These natives are very cleanly in their habits; but life passes with them in a very luxuriously lazy manner. They have wells for special purposes,—for bathing, for cooking, for washing; and on places where only hot vapour escapes from the ground, they have regular vapour baths. Upon the heated ground they have constructed warm houses for the winter season, in which no vermin of any kind is able to exist; and in the middle of the settlement stone slabs have been laid down, which receive and retain the heat of the ground in which they are sunk. This is the favourite lounge; and here, at any hour of the day, but especially when the shades of evening are closing round, all the rank and fashion of Ohinemutu may be seen, wrapped in their blankets, luxuriously reclining on the warm stones. Yet although the whole atmosphere in and about Ohinemutu is constantly impregnated with watery vapours and sulphureous gases, as to make them plainly perceptible to the sense of smell (Anthony Trollope in his vernacular calls it '*putrid stench*'), this seems only to improve the physical condition of the inhabitants, for they are known to be a robust set of Maoris, and they declare that this atmosphere is by no means unhealthy, although they live in a perpetual cloud of steam. Mosquitoes and sand-flies share the same fate as the other vermin.

Lieutenant Meade was quite enraptured with this luxurious sort of existence. He says:—' Old and young of both sexes meet in the lake every evening, nearly the whole population taking to the water, which is of an agreeable temperature, or like an ordinary warm bath. The lake seemed alive, for the rising steam prevented any more than a portion containing the bathers being visible; and the scene was a curious one. From every side were heard Maori songs and shouts from the players at some native game, and joyous peals of laughter came ringing along the surface of the water from beyond these misty veils. A number of the prettiest young girls in the settlement were seated in a circle in *very* shallow water, looking like mermaids, with the moon streaming over their well-shaped busts and raven locks, singing a wild song, and beating their breasts to the changing time with varied and graceful gestures. The choruses of the songs which followed were joined in by scores of voices; but ever and again even their voices were hushed and stilled, while, with a weird and rushing sound, the great geyser burst from the still water, rising white and silvery in the moonbeams which shone from the dark outline of the distant hills, and dashing its feathery spray high against the starry sky. The scene was the very incarnation of poetry of living and inanimate nature.'

This was the great geyser, which begins playing about the end of December annually, and gradually settles down quietly about February. Several tourists and visitors have been disappointed at not seeing this vast spring in its active state, and the only one who has given a description of it is the writer last quoted. He says:—'At first the eruption occurs with great regularity every twelve minutes, and lasts about twenty-five seconds. A vast column of boiling water, surrounded by glittering jets of spray and curling wreaths of steam, rises in one grand bouquet to the height of forty or fifty feet, an altitude which it retains for some seconds, and then slowly subsides into the bay whence it rose, and dies away in a surf of seething foam, leaving huge banks of steam rolling slowly up the dark hill side,—an exceedingly grand sight.'

First impressions of Ohinemutu are anything but favourable, for, go where you will, there is nothing but boiling springs, steam-holes, and hot mud to be met with. Some of these, used for cooking purposes, put one in mind of a boiling-down establishment. If the natives ever get drunk, there must occasionally be some most miraculous escapes, for the narrow paths, like Tennyson's 'Brook,' 'go in and out' among these springs and water-holes in such an intricate manner, that a step to the right or left would send one to eternity.

In some places, within a few feet of each other, may be seen a perfectly clear and pure boiling spring, in which are kits of potatoes and other food cooking, while a few feet off will be found good drinking cold spring water. Then, in close proximity to both, a hole in which the liquid mud is boiling and jumping up, making great pancakes.

Perhaps one of the most wonderful and interesting spots is the low point of land that runs into the lake from about the middle of the settlement, and on which once stood a Pa that long ago sank bodily into the lake. The carved tops of some of the large posts that formed the palisading of the Pa may still be seen; and in 1841, just forty years ago, when Colenso visited this place, he says: 'Ohinemutu is a large and fenced town on the banks of the lake celebrated for its boiling springs. The large spring at this place was boiling furiously, throwing out many gallons of water a minute, which rolled away steaming and smoking into the lake, a second Phlegethon. The sulphureous stench here is almost insupportable.' The blade of a knife immersed for a short period in these waters soon becomes as it were superficially bronzed. The natives who live in this neighbourhood are, when travelling, easily recognised as belonging to this district, in consequence of their front teeth decaying, unlike those of other Maoris. The natives of this village are celebrated among other things for their manufacture of tobacco pipes. These are carved out of a white stone which is found in the neighbourhood.

It is said that nowhere else in New Zealand can carvings in such profusion be seen, some of them being very old, each panel having the Moko of some Maori ancestor, the whole forming a carved history.

Colenso, during his first visit to this place in 1836, says : ' Nowhere in New Zealand have I seen anything that could be regarded as an idol, although some persons have said that such exist. This absence of all carved gods among the New Zealanders offered to me a very attractive trait in their national character. They are too much the children of nature, and perhaps too intellectual, to adore wooden images or animals ; and I often heard the natives deride the pewter images of the Holy Virgin which the Roman Catholic priests have brought into the country. What a noble material to work with, for the purpose of leading them towards civilisation !' John White says that some of the Maoris brought one or more gods from Hawaiki with them. The famous ones were brought by Kui Wai and Hangaroa. They were five in number, two of which, called Ihungaru and Itupaoa, remained to modern times. The Ihungaru was formed of a lock of human hair, twisted with a rope of Aute (the paper mulberry bark), and was kept in a house made of wood brought from Hawaiki for the purpose, and thatched with Mangamanga (a crêeping fern). This fell into the hands of Hongi and the Ngapuhi tribe at the storming of the Mokoia Pa in Rotorua in the year 1818, and, being carried from the little islet where the fortress stood, was brought to an eminence overlooking the lake, and, through the spirit of revenge, there cut to pieces with the tomahawk of the victorious Ngapuhi.

THE MAORI LEGEND OF HINEMOA.

In the middle of Lake Rotorua is the island Mokoia, a conical hill rising about 400 feet above the level of the lake, and with a Pa on the highest part. When visiting this interesting place, we went up through amongst the cultivations to the hill top where this great Pa stood. It was taken by the Ngapuhi natives in 1818, and has never since been fortified. The ditches and burial pits are undisturbed, as cultivation is not allowed to encroach on these. The place is Tapu, as the last stronghold of the great Arawa tribe of Rotorua, the ancient Ngati-whaka-ane.

This island has always been thickly inhabited, and is carefully cultivated. It is best known as the scene of one of the most celebrated Maori legends, that of Hinemoa, the ancestress of the present inhabitants of the island and of the district round Rotorua, including the village of Ohinemutu. She was a chief's daughter, a great beauty, but very headstrong. Finding her family opposed to the object of her affections on account of the inferiority of his birth, she planned an elopement by swimming across the lake, supporting herself when tired by a string of gourds round her neck. On reaching the island quite exhausted, she concealed herself in a warm bath, till her lover found her hiding behind the rocks, when, throwing his garment over her, 'she rose from the water beautiful as a wild hawk, and stepped on the ledge of the bath graceful as a shy white crane.' He took her to his home, and she became his wife, and they lived together ever after. The bath is most carefully preserved by the natives, and they show it to strangers with great pride and pleasure. It is indeed a beautiful spot on a beautiful island. A few boulder stones crop out of the water of the shallow border of the lake, covered with a profusion of flowering shrubs and plants. The ground in the neighbourhood is carefully cultivated, and the landing place just alongside the bath is a pretty, white, clean sand beach. The bath, heated from a boiling spring, is of a most delicious temperature. The tourist will visit the spot where the beautiful female came out of the bath, and fell into the arms of her adopted. The following is the story, as the Maori tells it, in all its purity and simplicity :—

Tutanekai was the foster son of Whakane, who lived on the island of Mokoia, in the lake of Rotorua. Whakane had several sons and one daughter; but Tutanekai was a foster son, the son of a stranger. His foster father treated him as kindly as if he were his own son. At a distance, on the opposite shore of the lake, was another village of the tribe, called

Legend of Hinemoa.

O-whata, and there resided the beautiful girl Hinemoa, the daughter of Umukaria, the chief of that place. (*Note.*—This village is about a mile beyond Sulphur Point, on a jutting promontory very little elevated above the lake; a large boulder stone, near the native fence, is shown to strangers as the place where the misguided young woman, before she took to the water, laid her Kakaha (outer garment), so that it might be seen by her bereaved relations early next morning.) At this time the sons of Whakane had grown up to man's estate, and the fame of the beauty of Hinemoa was everywhere heard, and the sons of Whakane were sick for love of that beautiful girl. So Tutanekai took a flute, and his friend Tiki another, and they went in company to Kaiweka, and in the calm summer night they played on them from an elevation above the lake, and the music was borne by the soft night breezes far over the waters to O-whata, to the home of the beautiful sister of Wahiao. And when Hinemoa heard the sweet sound of the flutes of Tutanekai and his friend Tiki, her heart danced within her, as night after night the floating music came. And she began to think it was the flute of Tutanekai which she heard, for they had often met before this at the meetings of the tribe, but had not yet spoken. At these meetings their eyes had sought each other, for they loved in secret; but Tutanekai thought to himself, 'Now, if I approach Hinemoa to pay court to her, perhaps she will be displeased with me.' And Hinemoa said to herself, 'Now, if I send a messenger to Tutanekai, perhaps he will not care for me.' At last, after many meetings at which their eyes only had spoken, Tutanekai sent a messenger to Hinemoa; and when she had seen the messenger, she cried joyfully, 'Ah, our minds are the same!' After this they were separated, and had no means of communication for some time. Now it happened on a time when the family and sons of Whakane were assembled at a merry-meeting, that the foster brothers of Tutanekai said amongst themselves, 'Who of us all have paid their addresses to Hinemoa.' And every one said, 'I have.' But when the question was put to Tutanekai, he said, 'I have, before any other.' His brothers replied, 'It is false; she would not listen to you, a stranger, a person of no consideration.' Then Tutanekai desired his foster father to bear in mind those taunting words which had been addressed to him, for he had seen Hinemoa, and they had agreed together that she should leave her home and fly to him; and she had asked him, 'By what sign shall I know that you expect me?' and he had answered, 'When you hear a flute play at night, 'tis I, come in your canoe.' So Whakane believed the words of his foster son, and bore them in mind. Again, at night, the friends have taken their flutes, and they are heard by Hinemoa far across the waters of the lake; but she has no canoe wherewith to cross the wide water, for her people, suspecting her intentions, had drawn them high on shore. 'O how,' she impatiently exclaims, 'how may I cross to Mokoia?' Despairingly she thinks she must remain, when across the water come the strains of the flute of Tutanekai. Oh! it was like an earthquake shaking and agitating that young girl with irresistible persuasion to fly to the disturber of her heart, her mind is confused, and she thinks at last, 'Would it be possible to swim?' Now she takes six hollow gourds and fastens them to her body to buoy her up, three on either side. Then she stands on a rock. The night is dark, and cold that inland lake ; her heart is trembling, but the flute plays on ; now she is on the shore at Wairerewa. She leaves her garments, and sinks into the waters of the lake. She swims towards the island, till, being exhausted, she drifts, buoyed up by the hollow gourds, on the rippling waves of the lake. Having gained fresh strength, Hinemoa swims again. The dark and dreary waters are wide, no land can she see ; her only guide is the strains of the flute leading her on her way, and at last she lands at Waikimihia, close to the residence of Tutanekai. Where Hinemoa landed there were hot springs, and she went into the hot water to warm herself, for she trembled with cold from having swam in the night across the lake of Rotorua ; and she also trembled with shame when she thought of Tutanekai. So, as Hinemoa was warming herself in the hot bath, Tutanekai said to his slave, 'Go, fetch me some water.' So the slave went and filled a gourd with water from the lake, close to where Hinemoa was lying. So she disguised her voice, and called out hoarsely like a man, 'For whom is that water?' And the slave answered, 'It is for Tutanekai.' Then said Hinemoa, 'Give it to me.' So he gave her the water vessel; and when she had drunk from it she broke it in pieces. Then said the slave, 'Wherefore break you the water vessel of Tutanekai?' but Hinemoa answered not. So he returned; and his master said to him, 'Where is the water?' Then the slave answered, 'The vessel has been broken.' 'By whom?' The slave answered, 'By a man.' Then said Tutanekai, 'Go again.' So the slave took another vessel, and went a second time; and when he had filled it, Hinemoa called to him again, saying, 'For whom is that water?' So that miserable slave answered, 'For Tutanekai.' Then said Hinemoa, 'Give it me; I am thirsty again.' So the slave gave her the vessel; and when she had drunk she broke it in pieces, as she had the other. Then the slave returned to Tutanekai, who said, 'Where now is the water?' and the slave answered, 'It has been taken from me again.' 'By whom?' 'Indeed I know not; he is a stranger.' 'He knew the water was for me, wherefore then break the vessel? Ha! I am enraged at this insolence.' Then Tutanekai took his war club and his mat, and went to the side of the hot spring, and called out, 'Where is the man who has broken my water vessels?' And Hinemoa knew by the voice that it was the heartflutterer who called. So she hid herself behind the overhanging rocks of the warm bath. Indeed, there was but small sincerity in that hiding; it was more for shame than a wish to be undiscovered. So Tutanekai sought about in the side of the pond, and soon he found her crouching bashfully beneath the rocks. So he caught her hand, and cried, 'Ha! who is this?' and she answered, saying, 'It is I, O Tutanekai.' Then said he, 'Who are you?' and she answered again, ''Tis I—'tis I, Hinemoa.' 'Ah ha! then let us go to the house at once.' Then she rose from the water, and, graceful as a heron, stepped on shore. Then Tutanekai took off his mat and spread it before her, and she took it and put it on; and the two went up to the house of Tutanekai, and, as was the custom of the olden times, in undisturbed slumbers they passed the night.

In the morning the people of the village arose, and came out as usual to cook their food. The day wore on, and they breakfasted; but Tutanekai came not forth from his house, so his foster father Whakane said, 'Tutanekai sleeps long: perhaps the boy is sick; go arouse him.' So one went and pushed back the little sliding door of the house, and looked in. 'What! what! four feet! Who can be his companion?' So he ran quickly back to Whakane, crying, 'I saw four feet in the house—four feet!' Then said Whakane, 'Who can be his companion? Go again.' So the man went, and, looking in, perceived it was

Hinemoa. Then he shouted, 'Hinemoa! Hinemoa! Tutanekai has got her! Hinemoa! Hinemoa!' So the tribe heard, and joined exultingly in the cry, 'Hinemoa! Hinemoa with Tutanekai!' So the foster brothers of Tutanekai heard the shouting, and they said, 'It is false;' but it was envy caused them to say this. Then out came Tutanekai from his house with Hinemoa by his side, and every one saw it was really her. Then the foster brothers said, 'So indeed it is true.' The descendants of Hinemoa and Tutanekai are living at Rotorua to the present day, and their constant theme is the beauty of their ancestress, and how she swam across the lake to Mokoia.

Sulphur Point.

Sulphur Point is a favourite place for invalids, it being but half a mile from the Ohinemutu hotels. It lies south-east along the shore of the lake. The first pool is a very deep hole, about twelve yards in diameter, filled with a bluish sort of water nearly at boiling point. The hole looks like an immense funnel, and a stone which we threw into the centre of it did not go out of sight for nearly a minute as it descended; the water is very transparent. A little nearer to the lake is the pond called 'White Sulphur Bath;' this is a pool about the same size as the one just passed, but it is filled with thick muddy water, in which gas bubbles are rising in all directions, giving it the appearance of being in a state of ebullition, though in reality quite cool. It is credited with some extraordinary cures. The gas rising from it is said to be the same as the well-known laughing gas, and has the same effect on some systems, so that the bather should always keep on the windward side. To the right, steam rises from a rotten-looking heap of black spongy substance, in which is a boiling spring that spurts up a miniature geyser to the height of a couple of feet. This feeds the pool known as the 'Pain Killer,' which is said to be extremely efficacious in acute rheumatism. Not far off is a loathsome-looking hole full of mud covered with scum, and about four feet deep. Another bubbling mud-hole some ten feet across is close by, and is also covered with scum. A little further round the shore is a large flat covered with silicious incrustations, in which are a number of sulphur springs of various degrees of temperature. They are all small, none exceeding a yard in diameter, and all make large deposits of sulphur in combination with saponaceous mud,—the sulphur greatly preponderating, so as to be in some almost pure. These are the 'Sulphur Cups.' Near the lake is a disgusting-looking hole, a yard in diameter, full of a bubbling mess the colour of coffee and the consistency of porridge, and smelling like something nasty. It is quite hot, and of some depth, and is known as the 'Coffee Pot.' Further still round the lake is another hard flat, on which are what are known as the 'Cream Cups.' They are little cups of smooth hard polished sulphur enamel, in the centre of each of which bubbles up a little jet of water, or rather gas and steam.

Whaka-rewa-rewa Springs.

'What sights you meet, and sounds of dread!
Calcareous caldrons, deep and large,
With geysers hissing to their marge;
Sulphureous fumes that spout and blow,
Column and cones of boiling snow,
And sable, lazy bubbling pools
Of spluttering mud that never cools,
With jets of steam through narrow vents
Uproaring, maddening to the sky,
Like cannon-mouth that shout on high,
In unremitting loud discharge,
Their inexhaustible contents;
While oft beneath the trembling ground
Rumbles a drear persistent sound,
Like ponderous engines infinite working
At some tremendous task below.'—*Domett*, C. 1, p. 7.

Whaka-rewa-rewa lies two miles from Ohinemutu on the east bank of the Puarenga creek. As it is a little off the direct road to Tarawera, and being so near to the Ohinemutu springs, it is often passed unobserved by travellers; but the springs here exceed those of Ohinemutu in variety and extent. Seven or eight of them are periodical geysers, and the natives say that they play alternately, and during strong easterly gales will sometimes play all together. The number of smaller springs, boiling mud-basins, mud-cones, and solfataras scattered over about one and a half miles along the banks of the Puarenga river to Arikiroa bay must be counted by hundreds; and what seems strange with natives who have no written language, almost every one has a native name descriptive of its character, or of some incident connected with it.

The first lot of springs are some highly medicinal baths, and a little further, on the left, is a large, hideous-looking hole, full of mud boiling furiously, and still further yet another. No words can paint the horrible appearance of these holes, the mud in them, dark blue-grey in colour, boils and sputters and hisses in a most vicious-looking manner; huge bubbles are formed, which burst with an unctuous sound, spurting up the slimy mud to the height of several feet. They are never quiescent, the weird commotion goes on continuously. Horrible and awe-inspiring as is their appearance, there is a strange fascination in them.

A little further is a large flat full of gurgling solfataras, with greasy-coloured water. Here are also a number of mud-cones, which are the largest in the whole Lake district, one being some six feet in height.

We now cross the river to Whaka-rewa-rewa proper, and, passing a few native Whares (huts), come upon a scene whose dismal weirdness could scarcely be exceeded. The Whares are at last shut out of sight, and no sign of life disturbs a picture of desolation positively appalling in its wildness. It would be vain to attempt a description of all the wonders of this dismal-looking place,—one or two must suffice.

The path leads through very rotten ground of red clay or decomposed rock, known to geologists as rhyolite, from which steam issues in all directions. To the right is an intermittent geyser of dirty grey incrustations, every cranny and fissure of which is lined with primrose crystals of sulphur, and with white irregular little stalactites, each one of which emits puffs of steam. On the top is a basin some ten or twelve feet across, in which the water is boiling violently, the whole body of it being occasionally raised a foot or two. Another geyser to the left, under the hill, is the Waikite, which issues from the top of a flat silicious cone over 100 feet in diameter, and 30 feet high. It rises between green Manuka and fern bushes, and presents an extremely picturesque sight. The cone is formed of the white silicious deposit, and has numerous fissures and crevices, and all are incrusted with sulphur crystals. The hot vapours issuing from these crevices and fissures smell of sublimated sulphur, not of acid or hydrogen. At intervals of eight minutes the spring throws out a column of water two or three feet thick to the height of six or eight feet in the winter time; but during the months of January and February it shows itself in its full glory,—spouting to a height of thirty or thirty-five feet!

The next large spring is called Pohutu, with a basin of about twelve feet wide. The masses of silicious deposits are here very extensive, and piled up to a height of more than twenty feet. The springs which supply the large bathing basins of over fifty feet in diameter—in which the natives, men and women, promiscuously bathe for hours, all comfortably smoking their pipes and chatting together—are called Parikohuru and Paratiatia. Not far from Pohutu will be noticed a perfect little geyser, throwing up a solid column of water to the height of nearly four feet. The water in the locality is evidently highly charged with silica, and articles left in it for a fortnight or so will be covered with a beautiful crust of alabaster whiteness.

On the west side, towards the peninsula, there is one other large basin, sixteen feet long and six feet wide, full of hot water, the temperature 185° Fahr.; and close alongside is a cold water basin, eighty feet long and fourteen feet wide, temperature only 55° Fahr. This basin contains yellowish-white water acidulated with sulphurous acid.

An invalid visitor who spent some weeks at this place, gives a very lucid description of the wonders of Whaka-rewa-rewa:—' As we approach the springs, there is nothing either grand or pretty about the scenery. The dull, heavy view is slightly relieved by the bright foliage of a few weeping willows, while the creek which rushes down between its high banks is the only thing that is romantic about the place. We are now amongst the strong-smelling steam, and already we hear the thud thud of the boiling mud, and the roar and noise as of mighty engines working in the bowels of the earth, from which the steam is rushing with great force. As we proceed, the ground is broken and full of deep holes. Suddenly we come upon one of great size, the bottom of which is a mass of boiling black mud of great depth. A few yards further is a large jet of steam, strongly impregnated with sulphur, and rising many feet from below the surface. On leaving this, we have holes to the right and left and ahead, all containing boiling black mud. After crossing a small stream, we come to a plateau of some substance not unlike dirty ice (silica thrown up by the water). This is full of holes, some of which are deep, but quite dry; and although the sides of the holes look very porous, others close by are seen to be full of water. In some of these places the water is warm, in others cold, and in many instances it stands at a different level in a most unaccountable manner. Away to the left of this is a warm lake, along the far side of which are several boiling springs of peculiarly soft water, strongly impregnated with alkalies. We next come upon one of the most remarkable wonders of the place. It is a very large and intensely hot spring of unknown depth, and the water of which is as clear as crystal. If a large stone is thrown into this spring, it can be seen sinking for a considerable time. A little further on may be seen a powerful spring, which throws a considerable body of water a few feet in height. The rocks in the vicinity of this spring are covered with a crystal crust, evidently formed by a silica sediment from the water. Continuing round to the right, we come upon three large hot springs, one of which, upon anything being thrown into it, commences boiling and hissing like an enormous body of soda-water. Lower down, and close to the creek,—the water of which is cold,—are several small boiling springs. We now take the track that runs to the back of these springs, and proceed over a small clay hillock, passing on our way two boiling slate-coloured mud-springs, one of which is very hot, and contains a vast amount of sulphur. From the top of this hillock a good view of all the springs is to be obtained. At its base are a lot of curiously-formed slabs of stone, all more or less covered with flowers of sulphur, and amongst which are numerous springs and innumerable jets of steam. The great peculiarity of these springs is the difference of the water flowing from each of them, as no two appear to produce water of a similar character. Some are quite clear, but taste different; some are of a greenish-yellow colour, some quite dirty, some acid; some taste of soda, many contain sulphur, and one is as salt as the sea; but all are boiling. Leaving these in the rear, and passing on towards what was the great beauty of Whaka-rewa-rewa, we pass on our left a very powerful mud-spring of intense heat. This spring sends the mud flying all round at least six feet. A mass of mud is first seen to rise like an inverted basin for about a foot. This then bursts, and is thrown out of the hole. The spring is very active. It is about five feet deep and ten in diameter, and as the whole mass is in constant motion, the reader may form some idea of the weight of mud in activity and the amount of power required to work the mass. We now come to the grand spring of the place, which rises out of a cone-shaped mound about fifty feet high, the sides of which are covered with a white silica sediment from the water, quite hard and beautiful. All the sticks, bushes, leaves, etc., that come within the reach of the waters of this spring get beautifully coated over.

'There is a strange intermitting spring here, which sometimes stops altogether. It will throw a column of water, a least four feet in diameter, in spurts of about half a minute's interval. The first would rise about ten feet, the second twenty feet, the third thirty feet, and the fourth and last of the series forty feet in height. It would then pause for two minutes, only to commence again with the same regularity and precision as before described. When not active, and between the spurts, you can look down this wonderful place; the noise resembles that of a hundred steam-engines at work. The natives use these springs for all kinds of skin diseases and rheumatic pains, having particular springs for each different ailment.

'A little higher up than the above eccentric spring, there is a very curious hole, surrounded by four slabs of stone about two feet high. In the good old times when fresh-cooked Maori was plentiful, this hole (for the steam is whizzing up from the bottom of it) was used as a pot for cooking the royal Maori dish of brain sauce. The old Maori, when describing the whole process, smacked his lips, and his eyes glistened with the recollection of something that the "pale-face" has banished from New Zealand.

'We now return again to the stream, and, following its course down over some very rotten ground, a district is entered which altogether defies description. Nothing is apparently to be seen, yet almost every step opens up new wonders. There is no beauty about them —far from it; one and all are hideous and disgusting beyond conception; but their hideousness is so fascinating, their variety so unexpected and startling, that one lingers among them, gazing down at their infernal glamour, unable to tear himself away. The ground is honeycombed, soft, and shaky, and the utmost care must be exercised in wandering deviously among the Manuka. Anything like enumeration is altogether out of the question. First, one suddenly finds himself on the bank of a huge hole, fifty feet across, full of the most beastly-looking coffee-coloured mud, from which the gas is escaping in rising bubbles over the whole surface; a step further, and one almost plunges into a tremendous caldron of unctuous boiling slate-coloured mud. Turning from this, the explorer is startled to find that he is looking into a large hole of white water, like the waves that beat against chalk cliffs; another step, and it is a bubbling pond of dirty grey water, covered with a villanous-looking oily scum. A little further, and more holes of coffee-coloured slime reveal themselves; and then a pond of light yellow water, round whose edge has settled an iridescent scum of greasy appearance; more mud caldrons, varied by a pond of dirty green, or a hole of reddish or sulphur-coloured water. And so on over acres of ground, while steam rises from cracks and holes in every direction, and a most villanous sulphurous smell pervades the air; and at last one is relieved to get away from these infernal regions, and shake off the glamour they have cast over him, and breathe again the fresh air of the desert.' Such is Whaka-rewa-rewa, in interest and weirdness second only to Rotomahana and Orakei-korako.

OHINEMUTU TO WAIROA, 9 MILES.

From Whaka-rewa-rewa the country is uninteresting until you reach the cool shade of the Moerangi bush, through which, for about two miles, the scenery is very pleasing, and on emerging from the shade of the old forest the traveller comes suddenly upon the prettiest lake in the whole district—

Lake Tikitapu.

This is a small sheet of intensely blue water about a mile long, of triangular shape. It is closed in between steep, partly-wooded heights. The hills rising from its northern and western banks, clothed with magnificent forests, cast a strong shadow on its deep blue waters. Such a change, and so sudden, from the boiling water and mud fountains, with their strong smells, makes one fancy that he has just arrived in a different country altogether. This lake has no visible outlet. Several old Maori legends are recorded of this pretty lake, one of them being the Saint George and Taniwha story of New Zealand,— a great combat between Tu-whare-toa and a real dragon (Taniwha). It is faithfully recorded in our unpublished *Sacred Book of Maori*, how the monster was conquered after a tremendous engagement, and then sentenced to go to the bottom of the lake, there to remain for ever. When the wind comes sweeping down upon the water, raising white-fringed waves, the Maoris say the Taniwha is turning himself over.

ROTOKAKAHI.

This is a very pretty lake, quite a gem for a painter or photographer. It is completely shut in by precipitous mountains, covered with stunted green scrub. The beautiful little island of Motu-tawa, near the centre at the far end, is covered round the margin with a number of grand old Karaka trees, and in a few solitary huts are the straggling remains of a departing race. The edge of the lake next the road is fringed with numerous shrubs and trees.

The Wairoa river leaves the lake, and, crossing the road, runs through the old settlement of Kai-tereria. The road winds alongside the river among native and European trees and shrubs, past the hotels near the old mission station.

The plan at present for travellers visiting Rotomahana is to leave Ohinemutu after dinner (midday), arrive at Wairoa after a two hours' drive, make arrangements that afternoon with the natives for guides and canoes or whale boats, sleep at Wairoa that night, and start at sunrise next morning for the Terraces. It is advisable to start before the sun is well up in the heavens, for as the day advances the wind often rises and increases in strength, making it both unpleasant and dangerous to cross Tarawera.

Lake Tarawera.

Tarawera is perhaps the most extensive as well as the most beautiful of the lower lakes. The shores are much indented, and covered with trees, whose overhanging foliage droops to the water. On the eastern side, from a chain of densely-wooded hills, rises a very remarkable and lofty mountain, whose summit is shaped like a truncated cone, but with symmetrical sides and upper edge. This is Mount Tarawera, a huge bare mass of rock 2000 feet high, —a regular table mountain. The natives have many traditional stories about it. From east to west, Tarawera lake is about seven miles long, with a breadth of about five miles. The lake is considered to be very deep, as its shores are mostly rugged rocky bluffs, shaded with the Pohutukawa trees, and for extent of scenery surpasses in wildness and grandeur all the other lakes. The outlet of the lake is the Tarawera river on the east side, which flows into the sea at Matata. Besides numerous small streams, Tarawera receives the discharge of five small lakes,—from the south-east, the joint discharge of Rotomahana and Rotomakariri; from the north-west, the waters of Okataina and Okareka lakes; and from the west, the Wairoa river, which, flowing from Rotokakahi, at a short distance from the mission station, forms a picturesque waterfall eighty feet high, and empties into the lake through a narrow gorge of rocks. At the Wairoa, besides the hotels and a few deserted huts, the settlement consists of church, school, and schoolmaster's house; but the church is without a congregation, the school without pupils, and the master's house is tenantless. The war scattered the flock, and the shepherd just saved his life. His next-door neighbour, Volkner, was hanged.

We are now in sight of the great Tarawera lake, and have also a splendid view of the sacred Tarawera mountain. It is sacred from being the principal burying-place of the powerful Arawa tribe. It remains Tapu to this day, consequently no person is allowed to ascend it. Our attention is attracted by a noise to the right, and a glimpse is caught of the Wairoa creek, as it glistens among the bushes. The waterfall here is not of a great height, but it makes a succession of falls as it descends from the top of the cliff to the shores of the lake, a distance of 1083 feet. The track from the hotel to the border of the lake is down this steep and dangerous cliff.

F

All this is, however, fully compensated for by the beauty of the lake and of the scenery around it. Kariri Pa is built upon a strip of land running into the lake. The spot is well adapted, the approach to it from the mainland being very narrow, and this has been at one time made very secure by a strong double row of palisading. As the land stretches into the lake it increases in width, especially on the right hand side, where it forms a splendid boat harbour. From this point a cliff rises, and forms a high perpendicular wall for nearly three sides of the Pa, thus making the defence of the place comparatively easy.

A sail in a canoe in fine weather is enjoyable in any part of the world. But in a great Waka, or New Zealand war-canoe, with a female crew, a fine fresh breeze blowing, just enough to create a sparkling ripple on the surface of the lake, it is something to be remembered. Starting from the Wairoa, this arm of Lake Tarawera is without exception, in point of pure picturesque lake scenery, the most perfect and exquisite of any in the Lake district. The arm is quite narrow, and is enclosed by high, bold hills clothed to the water's edge in a luxuriant robe of noble trees, charming in their varied tints of green and purple, and graceful in the rounded billowy masses of their foliage. No bit of scenery is to be met with anywhere in which is to be seen in greater perfection the one only beauty to be found in New

Zealand landscapes,—bush-clothed ravines; and the eye dwells with pleasure on the many-blended tints, from the delicate pea-green of the tall fairy-like reeds that wave rustlingly along the water's edge, to the sombre purple-green of the rolling clumps of giant Totaras. The clear deep water reflects the whole in pictures of dazzling brightness, and sends back the pale light of the passing fleecy clouds in flashes of silvery whiteness, brought into strong relief by the dark gloom of the sunless side of the deep bay, while the brown reed roofs of the boat sheds serve to break the monotony of the coast line.

As the bay is left, two pretty low-wooded bluffs reveal themselves on the left, appearing at first like islands, and then the lake widens out, and the eastern shore recedes far away into a large curving bay. Then the bush is entirely left behind, and on the right rises sheer from the water's edge a chain of steep fern-covered hills, prettily broken on their summit into miniature peaks, sharp and well defined ; many pretty bits of perpendicular cliffs of solid rock rise straight up from the water, shaded by overhanging clumps of red-flowered Pohutukawa trees. Far away on the left rises the dull mass of Tarawera mountain, reddish in colour, and bare, bleak, and barren in appearance,—a most dismal-looking object. Somewhere about half-way down the lake, a little cluster of Whares, surrounded with maize gardens, will be seen in a small bay. Farther along, the Pohutukawa trees line the entire shore, hanging over the water in graceful festoons. About this place, a large stone will be pointed out, standing about four feet out of the water. It is said to be the remains of a great Taniwha that causes the water to be troubled. In passing, it will please the Maoris very much if they are asked to pull near to the stone to enable you to put on it a small piece of money, or in fact anything. They say it will appease the monster, and keep the water quiet until you return.

When fairly out in the Tarawera lake, and surrounded with such grand and romantic scenery as meets you at the rounding of every headland, you will believe that this is by far the prettiest of all the lakes. It is amusing to watch the paddles going merrily to the music of the Maoris' song, with the wild refrain, 'Waka-tana Kea-wheta Haka-tu,' the 'tu' coming out with a great grunt and a strong pull altogether, the song and the paddling keeping beautiful time.

In drawing near the end of the voyage, the water of the lake, although very deep, gets gradually warm and warmer as you get near to the Kai-Waka creek, where the water is not only warm, but decidedly hot.

A pleasant walk of about a mile alongside this hot-water creek will bring you to Rotomahana—

> 'Not far from where, 'mid reed and sedge,
> The warm Mahana's rapid tide,
> A mile-long stream scarce six feet wide,
> Comes rushing through the open pass
> To Tarawera's ample lake.'—*Domett*, C. 18, p. 269.

ROTOMAHANA.

> 'And is this Yarrow?—this the stream
> Of which my fancy cherished
> So faithfully a waking dream,
> An image that hath perished!'

The first sight of this lake is disappointing, more particularly to those visitors who have been picturing to themselves all the pretty and grand things they have seen either sleeping or waking during a whole lifetime.

About the lake itself there is nothing grand or wonderful, nothing to be seen at first sight but 'a small dirty green lake, with marshy shores and desolate and dreary-looking treeless hills about it.' And you ask, inwardly of course, Where are all those unique and marvellous wonders of nature we have travelled so far to see? You look at your guide with disappointment pictured in your face, and he quietly says, 'Taihoa' ('Wait a bit'); and, starting off along the narrow path leading to the lake of which you have just come in sight, he calls, 'Haere mai' ('Come along'). You hold your tongue and follow the Maori.

So much has been written, and so well written,

descriptive of the wonders of Rotomahana by occasional visitors, that, as we cannot do better, we will select a few passages from their notes, making the description more intelligible and complete where to us it may seem imperfect.

The lake is small, and, comparing it with Tarawera, which we have just left, the scenery is bald and bleak, and not at all inviting on first acquaintance; but when seen, as it ought to be, in the early morning before the sun bursts out in the east, dispersing the clouds of steaming vapour, or towards evening, just before he disappears in the west, something more than a dirty green lake will be discovered.

Let the reader imagine a deep lake of a blue-green colour, surrounded by verdant hills; in the lake several islets showing the bare rocks, others covered with shrubs, while from all of them steam issues from a hundred openings between the green foliage, without impairing its freshness. On one side a flight of broad steps of the colour of white marble, with a rosy tint, and a cascade of boiling water falling over into the lake. This cascade is formed by the silicious deposits, and the steps are about fifty in number, from one to two feet broad. The steps are firm as porcelain, and with a tinge like carmine. The concretions assume interesting forms of mammilary stalagmites of the colour of milk-white chalcedony. It is quite true that the lake lacks all and every beauty of landscape scenery, but it possesses other attractions which make it the most remarkable of all New Zealand lakes. It must be examined quite closely, as from the eye of the traveller on his first approach it is hidden, and only by the steam-clouds ascending everywhere is he led to expect something worth seeing.

Rotomahana is not quite a mile long from north to south, and only about a quarter of a mile wide. It is 834 feet above sea-level. The quantity of boiling water issuing from the ground both on the shores and at the bottom of the lake is truly astonishing. The whole lake is heated by it; and, on looking closely at the water, it is muddy, turbid, and of a smutty-green colour. Neither fish nor shell-fish live in it, but various kinds of ducks, water-hens, the magnificent Pukeko (*Porphyrio melanotus*), and the graceful oyster-catcher Torea (*Hæmatopus longirostris*) enliven the surface of the water. There are two little islands in the lake; one of them, called Puai, is a rocky cliff 12 feet high, 250 feet long, and about 100 feet wide. Manuka grass and fern grow upon it. The natives have erected on it, for the use of occasional visitors, small Ranpo huts. Hochstetter spent one night on this curious spot. His experience is unique. 'It is almost the same as living in an active volcano. There is a continual seething, hissing, roaring, and boiling all round about; the whole ground is warm. The first night, the ground upon which I was lying grew gradually so warm from below, despite the thick underlayer of fern and the woollen blankets that composed my bed, that I started from my couch, unable to bear it any longer. To examine the temperature, I formed with my stick a hole into the soft clay soil, and placed the thermometer into the aperture. It rose at once to boiling point. On taking it out, hot steam came hissing out, so that I hastened to stop the hole again. In reality, the island Puai is nothing but a torn and fissured rock, which, boiled entirely soft in the warm lake, threatens every minute to fall to pieces. Hot water bubbles up all around, partly above, partly below the surface of the lake. No fire is required here for cooking, for wherever we dug but a little into the ground, or cleared the existing crevices of the crusts formed on them, then we could cook our potatoes and meat by steam.'

Eastward of Puai, and separated from it by a channel only forty feet wide, is a second island, Pukura. It is of the same description as Puai, but smaller in circumference; it is higher by several feet, and has also some Maori huts on it.

At first the wonders of Rotomahana fell far short of *our* expectations. We were not struck with awe, nor was our breath taken away; and when our guides pointed out what they considered the principal features of the place, we at once gave expression to feelings of surprise and disappointment, for all we could see was a large cloud of steam, a few dingy-coloured banks, and some pools of discoloured water. It is for this reason that all visitors should perform the journey up the hot river Raiwaka in a canoe instead of walking the distance. Seen from the water, the first view is grand in the extreme. The first impression of anything is often lasting, and it is as well, when it can be done, to make that as favourable as possible. Not that the rule stands good for such a place as Rotomahana, for see it how or when you will, as long as it is thoroughly seen, there can be but one impression left on the mind,— that will be more easily imagined than described.

Again, those who come on foot have to cross and recross the creek, and, unless they prefer wading, this has to be done while perched upon the top of a native's shoulders. Once over the stream in safety, a few minutes suffice to bring us face to face with the Tarata fountain. To write an accurate account of the wonderful White and Pink Terraces, and one that would convey such an impression as would enable the reader to picture these places in his mind, would be impossible.

The White Terrace.

'A cataract carved in Parian stone,
Or any purer substance known,
Agate or milk-white chalcedon.
Its showering snow cascades appear,

> Long ranges bright of stalactite,
> And sparry frets and fringes white,
> Thick falling plenteous, tier o'er tier,
> Its crowding stairs.'—*Domett*, C. 17, p. 269.

At the east end of the lake, with its terraced marbled steps, is the first and most marvellous of the Rotomahana curiosities. About eighty feet above the lake, on the fern-clad slope of the hill, from which in various places hot vapours are escaping, there lies an immense boiling caldron in a crater-like excavation, with steep reddish sides thirty or forty feet high, and open only on the lake side towards the west. This is filled with perfectly clear transparent water, which, in the snow-white incrusted basin, appears of a beautiful blue, like the blue turquoise. At the margin of the basin the temperature is about 183° Fahr.; in the middle the water boils. One traveller gives a very felicitous description of what he calls 'a natural wonder which surpasses everything of the kind that has yet been discovered in this or any other part of the world.' Te Tarata flows from a furiously boiling pool, which fills a deep crater opening on the side of one of the mountains surrounding the lake. The sides of the lake are lofty and perpendicular, and its dark frowning walls afford a striking contrast to the huge towering columns of glistening white steam ever rushing upwards from its mouth. Hochstetter says: 'Te Tarata is a geyser the eruptions of which equal in grandeur the famous eruptions of the great geyser of Iceland. The New Zealand basin is larger than the Iceland one, and the mass of water thrown out is immense. The deposit of the water is, like that of Iceland, silicious, not calcareous. It has the appearance of a cataract plunged over natural shelves, which, as it falls, is suddenly turned into stone.'

The size of the crater, at the level where the violence of the central action forces the boiling waves over the lower margin of the pool, is about sixty by eighty feet. The water is of an intense and brilliant blue, the reflection of which slightly tinges part of the column of steam, but the action of the vapour in escaping keeps the middle of the pool perpetually raised in a cluster of foaming hillocks several feet above the general level. From the mouth of the crater the wide-spreading waters fall in thousands of cascades from terrace to terrace of these crystallized basins. The water from each successive pool escapes in little curving jets to fill more numerous and broader pools below, or falls in a curtain of glittering drops from the fringes of crystal and glassy stalactites which form the margin of all the basins and terraces, and finally flows into the Rotomahana over a smooth hard flooring of a semitransparent white glazed surface, which paves the shore of the lake for a considerable distance, and 'that water has been dripping over century after century, Nature all the time carving for herself these wonderful hanging ornaments and exquisite cornices with the prolific hand which never stints itself in space because of expense.'

> 'Each step becomes a terrace brow,
> Each terrace a wide basin, brimmed
> With water brilliant, yet in hue
> The tenderest delicate harebell blue,
> Deepening to violet.'—*Domett*, C. 17, p. 269.

The water in the several basins is of the same deep blue as at the source, but the crystal margin, as well as the delicate crystallized tracery (reminding one of lace in high relief), which covers the whole of the broad flight of steps and curving terraces, are as white as driven snow, except that in a few places, for the sake of contrast, a delicate pink hue is introduced. Anything so fairy-like in nature is seldom to be seen.

In shape, most of the terraces somewhat resemble the curved battlements of ancient castles, though not so lofty; and the margins of the pools which they contain are disposed in almost symmetrical curves, each of whose extremities rests on the swell of those adjoining.

The tourist may here select a swimming bath of any temperature, from a mild tepid in the basins nearest the lake, to a heat several degrees above boiling point at the crater. The depths of these pools vary from nine inches to as many feet.

The approach to the lower basins of the Tarata is very pretty and novel. It is, as it were, along a numerous set of narrow foot-walks, formed through the Manuka and other ornamental shrubs; the bottom of such walks being covered with smoothly-laid crystal tiles, along which is strewed choice specimens of incrusted sticks and boughs of plants of various colours, some resembling ice with a coating of fresh snow, while others have more the appearance of being pieces of moulded alabaster. The lower terraces, or perhaps, more properly speaking, the walls of the lower basins, are of a blue colour, resembling in appearance so much masonry composed of polished blue granite. The water contained in these basins is of a light sky-blue tinge, and has the effect of imparting a light mouldy colouring to the walls wherever it overflows them. The effect of these dark walls is very beautiful, and they are very much admired by all visitors. Rising as they do suddenly from out of an almost level and nearly white crystal ground, and being the only dark objects the eye can perceive, they form a grand relief and pleasing contrast to the prospect as seen from the edge of the lake. It is only when taking a general view of the terraces of Tarata from the base and near the lake, that its wondrous formation, stately grandeur, and its exquisite beauty are apparent to the tourist. It is, as it were, formed or built upon the side of a hill of considerable height, the site upon which it stands looking as if it had been dug out in a circular form, and then the whole space filled up level with the hill sides with a vast pile of magnificent masonry, perfect in workmanship, and composed of the rarest materials. At the sides, and from the level of the top of the highest terraces, may be seen the natural clay forming a wall along the sides and back of the great crater, from which flows the water supply that constantly ripples over the shining terraces. The space occupied by this great structure is about three hundred feet along the front of the base, the height being about a hundred and fifty feet. The face of the whole of the terraces takes a fine circular sweep to the front, the larger ones coming out from twenty to twenty-five feet upon the level; most of these contain pools of lovely blue-coloured water of different temperatures, those nearest the top being the warmest. They all form delightful baths,—in fact, the baths of the ancient Romans could not have surpassed them. Not only would the beautiful composition of which the sides and floors of these baths are formed throw into the shade the marble of the Roman baths, but the loveliness of the water, its softness and purity, its varieties of temperature, increasing so gradually with each rise, could not be produced by artificial means. The steps, too, are formed of a substance that would compare with moulded alabaster. Who would not bathe in such luxury as this? We have heard visitors remark that the only difficulty was to drag themselves out, and that they remained in the water so long, and ascended to such a high rate of temperature, that they were almost too weak to travel for some time after.

On ascending the terraces from the base, the dark-blue basins appear as if they had been expressly formed for gold and silver fish, and the snow-white

The White Terrace.

plateau running down to the lake seems to have been intended to teem with flowers and sweet-smelling shrubs and plants. As we proceed, the delicate beauties of the place are met with.

A large proportion of the pools in the terraces, more especially the smaller ones, are shaped in the form of gigantic shells. The water, as it continually ripples down the overhanging edge of the shell, forms stalactites of various colours, and many of them of great length. These have a really charming effect; and when seen in some of the larger cavities, with the sun shining full upon them, and the water rippling over and dripping from them, the scene is at once dazzling and beautiful. A curious effect is caused by the changing of the colour of the terraces, which are of a light colour at the sides, but become red as we advance to the centre, while at the top they are quite white. This effect is increased by the deepening of the colour of the water, which in the highest pools is of a rich sky-blue.

In numerous parts of the terraces, beautiful specimens of branches of shrubs, feathers, and even birds may be found most perfectly incrusted by the sediment from the water, which, upon evaporation, leaves a deposit that is perfectly hard, and gives, according to the particular part of the spring from which it is taken, the appearance of silver or ice or snow, or perhaps like that of flint, alabaster, or pure marble. Cloth, plaited flax, rope, etc., all get served alike. In fact, these remorseless waters keep adding daily to their stock. Should the water increase in bulk and run over, a fresh strip of ground soon becomes coated over, and in time is made 'beautiful for ever.' As we approach the top, the temperature of the crustation is felt in an unpleasant manner, even through strong boots; and the steam which now rises all around us gives warning that we shall soon be up to the boiling pool. We are now on the ledge of the top terrace, and are compelled to lower our heads to obtain a partial view of what is before us. The vast body of steam lies provokingly close to the water, protected from the wind by the great earth wall that surrounds the crater on three sides. A little patience, and we are rewarded. A puff of wind comes to our relief, the steam is carried off gradually, and at last we see the surface of the boiling water. It is at least one hundred feet in diameter, and nearly a true circle. A narrow ledge runs round some distance upon the level of the water, but he would be bold who would venture to follow along it. It is impossible for us to estimate the depth of the water; but the natives informed us that this wonderful hole is at times nearly dry, and that a man can scramble a considerable distance down the side of it. They say that the decrease in the quantity of water only happens occasionally, and is caused, they think, by the prevalence of wind from a particular quarter. At the far side of the crater,

The White Terrace.

and almost close to the steep wall, are five very powerful boiling springs, throwing large bodies of water to a great height. It would be impossible to speak as to the exact height, owing to the density of the steam. The power of these springs may be guessed at, when we take into consideration that, according to the natives, the water in the crater sometimes receded to a depth of at least fifty feet. Although warned by the natives that they did not consider it safe, our party managed to climb up to the top of the hill that immediately overlooks the crater. We found it soft in places, and extremely hot, especially to our feet, which at times would go more than ankle-deep in the earth, and when withdrawn small jets of steam would come from the place where the crust had been disturbed.

The view to be obtained from the summit quite repays the inconvenience. The view of the crater and terraces alone would repay it; but when it is known that from this exalted position the whole of the Rotomahana lake can be seen, together with the mountains of Kakaramea and Maunga-onga-onga, each nearly three thousand feet high, the reader may judge if it be worth while to have a little exciting climb or not when visiting the place. While we were descending, we had a full view of all the terraces, with the pools of coloured water most gloriously spread out before us. It was one of the grandest and most wonderful sights that can be imagined. No description, photograph, or picture could give a full and correct idea of its beauty.

One tourist says: 'How shall I attempt to describe Rotomahana? If photography cannot do it justice, which it does not, how can the effete tracings of my pen soar to the conception of the most remarkable and most beautiful phenomena of nature that I have ever as yet seen in the world! Niagara, with all its grandeur, is after all but a gigantic waterfall; and the Yosemite Valley is nothing more than a deep gorge, unequalled by any other gorge for wildness of beauty: but the wonders of Rotomahana stand, as far as I know, by themselves, not only unequalled, but incomparable.'

A lady visitor says: 'All of a sudden the whole terrace came in view, with the water flowing down its marble-like steps, and a fierce steam breaking out of the top basin. The weather was rather gloomy, everything was quiet round about, and Lake Rotomahana looked rather insignificant; but what did it matter compared to the majesty of this White Terrace, the first aspect of which was to me overpowering! Let us ascend the fairy staircase to the giant's castle. Any attempt at enumerating the steps would fail, for they are by thousands, of all sizes, from the segment of a cheese-plate to that of a twenty-ton galleon. Water is found to be trickling over the whole surface, to the depth of from an eighth to a quarter of an inch, at first almost cold, but gradually warmer, until it gets unpleasantly hot. Each step, large and small, is now seen to be a little reservoir of itself, being hollowed out on the top, and having a rim of rough rounded incrustation, the water in all being of an opaque turquoise-blue colour, increasing in intensity as the reservoirs get larger towards the top.'

THE EAST SIDE OF LAKE ROTOMAHANA.

> ' Through shuddering rocks, blanched ashy pale,
> Hot water, steam, and sulphur smoke,
> Commingling in one column dense
> Of white terrific turbulence !
> But other gentler feelings woke
> Its sister fountain, welling nigh,
> Whose bursts of grief for moments brief,
> Long intervalled, in stream outbroke,
> And then would sink away and die
> With such soft moan, relapsing slow
> Such long-drawn breath of utter woe,
> It well became its mournful name,—
> Koingo—Love's despairing Sigh.'—*Domett*, C. 17, p. 276.

From the lower terraces of Te Tarata a footpath leads along the slope of the hill to the first great fountain.

THE GREAT NGA-HAPU

Is reached after scrambling through a dense thicket of shrubs, close to the margin of, but about ten feet above, the lake ; the dense body of steam continually ascending from it indicates the locality. The basin is oval, forty feet long by thirty wide; the water is clear and transparent, and nearly always in a terrific state of ebullition. It is an intermitting spring; for a few moments calm, it then madly boils up, first on one side, then on the other, raising its foaming crest to a height of eight to ten feet, when a great surf of boiling hot waves lash the walls of the basin with wild uproar, so that the traveller, stricken with awe, shrinks back from the scene. The water does not overflow, but the natives have constructed an outlet which leads it to several bathing-places below. In the basin, the thermometer rises to 210° Fahr. It is, without exception, the most powerful and troubled spring we have ever seen. It is in constant motion, and is always throwing up water in some part or another. At short intervals it affords a grand display, which becomes most awful when at its height. It commences by lifting a vast body of water suddenly to a height of six feet from the back of the pool; then the whole of the surface becomes agitated in good earnest, until the whole body is in a state of violent activity, each part trying to outvie the other. The centre is always the best, and appeared to throw enormous bodies of water with wonderful rapidity to the height of fully thirty feet. When our party came to this spring it was quiet, and so they passed; but I remained to examine some delicate ferns, within about six feet of the basin, when all at once there was a roar, the whole body of water seemed to have been shot up, and there it was, away above me; then in an instant the steam-cloud burst out, and I found myself in a hot steam bath, and when it all subsided, I found myself wet to the skin.

The Great Steam Fog-horn.

The attention of visitors is generally so much engrossed with the wonderful evolutions of the great Nga-hapu, that they often pass along without noticing the great steam-pipe, which is situated a short distance up the hill side above the spring. This marvellous steam-pipe is really a hole in the hill side of about twelve inches diameter, from which the steam is constantly rushing with such violence, that the only matter of surprise is that the whole hill side is not blown out. There never seems to be the slightest indication of a stop; and as to the amount of pressure or the weight of steam that escapes from this hole, no idea could be formed. Many attempts were made to throw lumps of wood across or down it, but none either remained in or near it. Stones were then tried, but without effect; any that fell into it were at once thrown up, and in some instances to a great height. None of the party could be induced to go near enough to try to hold the end of a stick over it, or push some large-sized stones that were pretty close to the hole, over the mouth.

Te Takapo Fountains.

A little farther along the shore lies Takapo, one basin ten feet by eight, with clear, gently boiling water, the temperature 206° Fahr. This fountain, at certain seasons, rises in a jet thirty to forty feet high. There are other five or six, two being very strong and very active. It is curious that, although these springs are close together, each has water peculiar to itself, no two being alike. One spring has the power of coating everything that comes within its reach with a material resembling bronze; one is too bitter to drink; while another is considered by the natives to be the best water obtainable for culinary purposes. The natives have at one time been very industrious in the neighbourhood of these springs. There are some excellent baths constructed with large slabs of stone, and the waters that are led into them are considered by the natives to have a very beneficial effect upon a debilitated constitution.

The Wai-Kanapanapa Lower Springs.

A little farther onward is a small greasy-looking lake of yellowish-green water, deep, cold, and clear. It is remarkable only on account of its extraordinary colour; the name means dark and dismal-like deep water. The hollow in which this pool is situated is full of springs of boiling mud, probably not less than a hundred of them constantly playing in all directions at short intervals, all over the flat. This sight, grand as it may be in itself, is greatly enhanced by the fact that the hill side on the left of the gulley is covered with small holes which constantly send forth jets of steam. The whole range would appear to be neither more nor less than a hotbed of steam. The mud-springs in the gulley are not without their peculiarities. Those first met with are more liquid than the rest, and are much larger and stronger, throwing their black slime four or five feet high. They are also much hotter, and the stuff they vomit is of a very different nature to the rest. The majority of these springs throw up a stiffish grey-coloured greasy clay, which the Maoris call Kaikai, or food, and they eat it with great avidity. One lad who accompanied us, in spite of his having only just demolished a fair share, even for a native, of bread and meat, set to at once with quite a zest, and before leaving the gully had consumed fully a pound and a half weight; it was a fair double handful of the stuff. I was prevailed upon to taste it, and although it was not remarkable for flavour, it seemed palatable to the taste, and very clean for mud. It was infinitely preferable as a food to a great number of the messes that the Maori eats. The clay, as thrown up by these springs, forms a conical mound around the holes, varying in height from one to three feet.

THE KUAKIWI, THE NGAWHANA, THE KOINGO, AND THE WHATAPOHU FOUNTAINS.

A short distance past the native settlement lies Kuakiwi, a basin sixteen feet by twelve, filled with clear water, the temperature 210° Fahr.; and on the rise of the hill behind is the Ngawhana, a quiet hot-water basin; and close alongside is the sighing Koingo intermittent fountain, from which discharges of water take place three or four times a day, alternating with the adjoining one, Whatapohu, a basin nine feet by five, with temperature 202° Fahr. A few yards farther south there is what has been described as a curiosity, being

A Fountain, Solfatara, and Fumarole, all three in one.

It proceeds from a deep shaft-like aperture between ash-coloured rocks, as from a steam-boiler, hot steam and sulphurous gas issuing with a dismal moaning sound, and occasionally it throws out spouts of water.

TE KAPITI FOUNTAIN.

This is in a constant state of activity, and the sediment left by the water is extremely hard, and resembles glass.

ROTO POUNAMU, OR THE GREEN LAKE AND SPRINGS.

This Ngawha is about half-way up the hill above the little Nga-hapu; it is almost hidden in the bush, but it is a pool not less than forty feet in diameter. At certain seasons, the water in the centre rises to a height of from eight to ten feet, driving the boiling waves over the rocky margin with a suddenness and fury which renders both caution and agility necessary in approaching it. The red porcelain pavement from this pool extends to the lake, whose shores and surface are so covered with floating and stranded pumice-stone, that it is difficult to distinguish the outline of *terra firma* till the floating pumice has actually given way beneath one's feet, and the traveller finds himself walking into the lake. Some of the earthenware pavement here is thinner and more brittle than a tea-cup.

The following by another visitor seems partly intended for the above springs, and partly for Te Nga-whana. The first part of the description places the springs on the top of a hill, but this is a mistake;

it then goes on to say:—'About half-way up the hill side is a small pool, through which the water from this spring flows. It is not over fifteen feet in diameter, but the bottom of it presents one of the most beautiful sights it ever fell to my lot to witness. It is covered with a most lovely-looking sponge-like moss, fully eight inches in depth; it is of the most delicate structure, and is entirely composed of the sediment from the water of the spring above.'

This is one of the greatest wonders of natural history that I ever met with, and one that would set the most thoughtless unbeliever thinking. This fragile moss appears to be growing, but cannot be so, for it is against all experience that life should exist in boiling water. The sediment seems to have been collected by some unaccountable agency amid the turbulent waters that rush down from the boiling spring, and to have been formed into this delicate moss-like substance.

As we proceed along the side of the Green Lake Pounamu, we observe other effects of this particular water. The most delicate things, such as fern leaves, flowers, feathers, or fine twigs, are by the agency of this water coated over, and appear to be converted into silver-like crystals. The spring itself is not constantly overflowing. It slowly rises from a considerable depth, and commences to overflow gradually. Then it seems to be seized with a fit, and at once throws the water up in large volumes to the height of three or four feet, and continues to do so for about five minutes. It then quickly subsides, and remains inactive for about ten minutes or a quarter of an hour, and then begins again the same old game.

TE RANGIPAKARU FOUNTAIN.

This is the last great spring on this side of the hill, and the only novelty about it worthy of notice is the great and unpleasant noise it makes, very much resembling a steam-whistle.

These are a few of the curiosities to be seen in skirting along the eastern shores of Rotomahana, but every few steps, every minute, brings some fresh wonder to view, differing entirely from the last. Here a group of little mud volcanoes in full and rather comical action; there a furious boiling pool, clear as crystal, with periodical geyser eruptions; or again a miniature cliff of pumice-stone and silica; now a basin of boiling mud of a dull white, then a pink one, and then again a black; here a little geyser; then a solfatara, with sulphureous fumes issuing from a yawning orifice incrusted with crystals of sulphur; or occasionally a fumarole, from whose crater escapes a few fitful wreaths of smoke; while from a thousand cracks and crevices in the many-hued and decomposed rocks, jets of steam hiss forth. There are about twenty-five large Ngawha, as the natives term the hot springs of the Tarata kind, scattered along the eastern side of the lake, and many hundred smaller ones. The fact is that a description of each and every one would be tame and wearisome to both writer and reader. Our object is simply to point out the particular or distinguishing features of each district as we proceed with the tourist on our voyage of exploration; beyond that we break down completely, for the number of curiosities crowded together over a space of less than half a mile can only be seen, not described.

Sitting near the margin of one of these great boiling caldrons, the mind naturally tries to realize the scene by comparison, and who can assist in this so well as Glorious John Milton? In his *Paradise Lost*, after the great meeting that was held in the infernal regions to appoint Satan a commissioner,—as Burns puts it, 'Ye cam' to Paradise incog.,'—his compeers, to amuse themselves during his absence, went on an exploring expedition through Hades:

> ' Through many a dark and dreary vale
> They passed, and many a region dolorous,
> O'er many a frozen, many a fiery Alp ;
> Rocks, caves, lakes, fens, bogs, dens, and shades of death,
> A universe of death, which God by curse
> Created evil, for evil only good ;
> Where all life dies, death lives, and nature breeds
> Perverse, all monstrous, all prodigious things,
> Abominable, unutterable, and worse
> Than fables yet have feigned or fear conceived,
> Gorgons, and hydras, and chimeras dire ; '

until at last they found the very counterpart to the scene before us:

> ' Along the banks
> Of four infernal rivers, that disgorge
> Into the burning lake their baleful streams :
> Abhorred Styx, the flood of deadly hate ;
> Sad Acheron, of sorrow black and deep ;
> Cocytus, named of lamentation loud
> Heard on the rueful stream ; fierce Phlegethon,
> Whose waves of torrent fire inflame with rage.'—
> *Paradise Lost*, Book ii.

Every part of the valley not occupied by the swamp rushes, is covered with a hard crystallized crust, as white as snow, and strewn with various objects similarly incrusted, so that it appears, before you reach it, to be a curious continuation of the lake, over whose frozen surface had swept a snowstorm. The brittleness of this crust and of the caking of baked clay makes it necessary to step very gingerly, and in some parts to place layers of brushwood to walk upon. Some of the water seems to have the power of fossilizing wood and similar substances, whilst others merely cover the object over which they flow with a hard white crust. These crystallized leaves and other objects of beautiful and fantastic shape are scattered about in great profusion.

We would strongly recommend the tourist not to run over this district, but, if possible, pitch his tent alongside one of the geysers, and spend at least one full week exploring the many and ever-varied phenomena which almost every step discloses.

A Sunset at Lake Rotomahana.

We will now return along the shore of the lake, till we reach Te Tarata again. One traveller, returning in the evening after spending the day exploring the locality we have just left, says: 'We were greeted by a sight which defies description, but will never be effaced from our memories. The sun was just setting behind the sombre western hills, above us were clouds, orange, golden, and purple, of unusually warm and brilliant tints even for a New Zealand sky, before us acres and acres of water-terraces, such as might belong to some giant's palace in fairyland. Every ray of the sinking sun was caught and broken into a thousand prismatic lines by the countless crystals that hung like lustres round the margins of the successive basins, or mingled in the blue waters within them with the gorgeous reflection of the glowing clouds above. Lower still, as a foil to this glorious picture, lay the dark waters of the calm lake, buried in the deep shade which the mountain cast eastward, and motionless, save when the still surface was ruffled by the teeming flocks of wild fowl. Beyond the lake, towering dark and sharp against the warm western sky, rose the grim mountain, Te Rangipakaru, with its great crater vomiting dense clouds of sulphureous vapours. The feelings which the spectacle brought forth may perhaps be imagined, but the sight itself was one which no pen could well describe.'

Such is the famous Tarata—the White Terrace—and its surroundings. The pure white of the silicious deposit in contrast with the blue of the water, the green of the surrounding vegetation, the intense red of the bare earth walls of what Hochstetter calls the 'water crater,' and the whirling clouds of steam, altogether present a scene unequalled in its kind.

THE PINK TERRACE, OR TU KAPUARANGI.

'The Fountain of the Clouded Sky,
Tu Kapuarangi, fitly styled,
It flings its steam so wide and high,
'Tis rosy rime they climb this time;
For floors and fringes, terrace piled
O'er terrace, glow with faint carmine,
As fashioned of carnelian fine,
As if continuous full from heaven
Some wide white avalanche, downward driven,
Came foaming out of sunset, stained
With sanguine hues it still retains.
Behind the topmost terrace, lo!
A vision like a lovely dream!—
A basin large, its further marge
And all its surface hid in steam,
That thinly driving o'er it flies,
Spreads level with the level plain
Of smoothest milk-white marble grain;
And all around its nearer brink
A border broad of delicate pink,
That melts to lemon yellow dyes
That whiteness, and with even hues,
Fair as a rainbow laid in snow.'—*Domett*, C. 17, p. 278.

The Pink Terrace is the crowning glory of the Lake district; everything else fades into insignificance as you feast your eyes on this magnificent structure,—it is so wonderful, so grand, and so gorgeous in workmanship. Examine it how you will, from a distance or with a microscope: you have the great terrace buttresses sufficiently large to form a bathing-place for an elephant, or a beautiful miniature tiny basin with the water just warm enough for a bath for a newly-born child.

Anything so exquisite as this Pink Terrace does not exist in nature. The artificial marble-like steps, fringed with shrubs, ascend from the lake; and the platform, sixty feet above the lake, has a square of one hundred feet each way, containing a number of basins three or four feet deep, full of transparent sky-blue water, with a temperature from 90° to 110° Fahr.; and in the background, shut in by half-naked walls tinged with various colours, red, white, and yellow, lies the great basin, a caldron forty or fifty feet in diameter. This terrace possesses beauties quite unique and peculiar to itself. The colours are more delicate, blending into one another like the reflection of the rainbow on the sparkling waters. To complete the picture, let the traveller indulge in a luxurious bath, and let him select a basin to his taste, for here he can, as usual, have it at any temperature. The bather undresses on a piece of dry rock a few yards distant, and is in his bath in half a minute, without the chance of hurting his feet, for it is one of the properties of the stone flooring which has here been formed, that it does not hurt. In the bath, when you strike your chest against it, it is soft to the touch; you press yourself against it, and it is smooth; you lie about upon it, and although it is firm, it gives to you; you plunge against the sides, driving the water over with your body, but you do not bruise yourself; you go from one bath to another, trying the warmth of each. The water trickles from the one above to the one below, coming from the vast boiling pool at the top, and the lower, therefore, are less hot than the higher.

From the terraces, as you lie in the water, you look down upon the lake which is close beneath you, and over upon the green broken hills which come down upon the lake on the other side. The scene from the bath in the Pink Terrace is by far the loveliest, and is luxuriously captivating, or, as Anthony Trollope *feelingly* puts it, 'it is a spot for intense sensual enjoyment.'

After bathing, the traveller will go to the top, where, from the platform already mentioned, he will see the fountain or caldron itself, the waters of which are constantly boiling. To see it comfortably, he must get to windward, to keep the steam out of his face. On descending, the beautiful steps, formed of the pure yellow sulphur set off by the fretted stonework under the curves of the rocks, where not exposed to the sunlight are perfectly green,—there huge masses bright salmon-coloured, and here and there delicate white fretwork. The lips and sides of the baths are tinged with that delicate pink hue which one is apt to connect with voluptuous enjoyment.

The baths are shell-like in shape, like vast open shells, the walls of which are concave, and the lips ornamented in a thousand forms. Four or five persons may sport in one of them, each without feeling the presence of the other. Very truly one traveller says, 'I have never heard of other bathing like this in the world.'

A very good description of this terrace is given by a visitor from one of our southern settlements, evidently a business man from the style of his narrative: 'After taking a parting view of Te Tarata terraces, we arrived at the creek (Kaiwaka), which we recrossed in the same style as on the previous day, and then wended our way to the Pink Spring, distant about a mile and a half. In visiting this spring, as we did by land, we arrived at the top first, and I am inclined to think that is the best way to visit these beautiful terraces. This one is perhaps best described by a comparison with Te Tarata. It is not so large as the latter, but about the same height; its steps are more regular, being perfectly level, and usually running with the greatest regularity from side to side. In colour and beauty, Tarata is far surpassed by it.

The Pink Terrace.

The lovely tint would rival in bloom the cheek of any fair country maid. Lucky would be the manufacturer of paste and powders who could once produce an article that would impart such a lovely blush to any lady's cheek. His fortune would be made, and his name become famous. Where lives the young lady who would not become a customer?'

The material that the terraces are formed of excels the finest Italian marble. It is soft and delicate to the touch, yet its surface looks and shines like glass. The steps are pure white at the top, and as we descend they gradually change their colour, until in the centre and towards the bottom, a rich, full, and yet delicate rose-pink prevails. At the sides of the middle tier, this crystal substance is striped and spotted with lemon and bright yellow. Like those of Tarata, they have their pools and baths, but, beautiful as they are, those of the Pink Terraces far excel them.

A lady visitor from Sydney, New South Wales, went to the Lake district entirely by herself, and on her return published a pleasing account of her trip in a Sydney newspaper. Speaking of this place, she says: 'This terrace looked lovely, for the sun appeared now and then showing off the delicate pink line of the basins in all its delicacy. In some parts the beauty of the steps is quite overwhelming. Steam also issues from the crater here, and the water that fills the basins is of an indescribable blue tint, beautiful in the extreme; when bathing in it, the limbs look like pure marble. I bathed in the middle basin, of beautiful shape, and deep enough to let the water reach up to my shoulders. In the Ti-tree scrub I undressed, and the bath was reached from thence very quickly. The water was so exquisitely soft and agreeably warm, that I could hardly persuade myself to leave it. Behind me the steam was ascending from the crater, and I looked down upon the middle basins, upon the lake, and the opposite shore. I was apparently quite alone, for the natives who had brought me up in the canoe had retired, and let me say this for their discretion, that they did not once attempt to look at me. I suppose they are so accustomed to see their own women scantily dressed, that it never enters their heads to be curious on this subject. Besides, why should they wish to see European live statues, when they themselves are so far superior in shape?'

The following highly poetical, but not by any means over-coloured picture, applies exactly to the Pink Terrace:—'The terraces at a distance appear the colour of ashes of roses, but near at hand show a metallic grey with pink and yellow margin of the utmost delicacy. Being constantly wet, the colours are brilliant beyond description. Sloping gently from the rim of the crater in successive terraces, it forms little bathing pools, with margins of silica the colour of silver; the cavities being of irregular shape, constantly full of hot water, and precipitating delicate coral-like beads of a slight saffron. These cavities are also fringed with porcelain rockwork around the edges, in meshes as delicate as the finest lace. The deposits are apparently as delicate as the down on the butterfly's wing, both in texture and colouring. Those who have seen the stage representation of Aladdin's cave, or any other gorgeous Christmas pantomime, can form an idea of the wonderful colouring, but not of the intricate frost-work of this fairy-like scene, growing up amid clouds of steam and showers of boiling water. One is in utter doubt as to the evidence of his own eyes, for the beauty of the scene takes away one's breath. It is overpowering, transcending the visions of a Moslem paradise.

'As seen through this marvellous play of colour, the decorations on the sides of the basins are lighted up with a wild weird beauty, which wafts one at once into the land of enchantment. All the brilliant feats of fairies and genii in the *Arabian Nights* are forgotten in the actual presence of such marvellous beauty; life becomes a privilege and a blessing, after one has seen and thoroughly felt its cunning skill.'

OHINEMUTU TO ORAKEI-KORAKO AND LAKE TAUPO BY THE PAEROA RANGE.

WE will presume that the traveller has returned from Rotomahana to Rotorua (Ohinemutu), from whence he will make a fresh start. As we have already said, there are two roads leading from Ohinemutu to Lake Taupo,—the mail-coach road by Horohoro on to Niho-o-te-kiore at the Waikato river, thence to Taupo; the other is a bridle-path, passing along the Paeroa valley to Orakei-korako, thence to Taupo.

The Paeroa range along its whole length is a huge mass of boiling pulp; so rotten and unstable is it that no one has yet dared to ascend its slopes. The traveller can only pass along the great valley close to the hill, and look upon the wondrous heaving mass of boiling mud, earth, and rock, so much disintegrated by the enormous force of high-pressure steam and internal heat, that in some places, by taking but one step outside the footpath, the consequence would be a terrible and instantaneous death. There is not, however, the slightest danger, if the footsteps of the Maori guide are carefully followed.

From Ohinemutu the road follows the line of telegraph posts, leading again through Whaka-rewa-rewa, then up the low fern-covered range behind; descending again, we cross over a long flat country enclosed by ranges of low hills, on top of which the igneous rocks crop out grey and weather-beaten. Two fine streams of water are passed, and then by a rather rough path another low range is come to, down which on the other side the road plunges into a dense growth of underwood, which completely shuts out the light of day. A long gloomy avenue now leads through the valley to a stream spanned by a very primitive bridge. We have now arrived at about the loveliest spot in the whole Lake district. It is nature in its natural state. The banks of the river are covered with luxuriant vegetation, and many rare tree-ferns peculiar to this locality are to be found here in perfection. An *al fresco* siesta and a mid-day meal for man and horse may be at last indulged in to perfection, amongst the pretty natural scenery. The little river, spanned by a rude bridge, looks tempting for a bath; but beware, for it is hot, much too hot for bathing. However, cold water may be found a few feet to the left of the bridge, where a delicious stream of cold water runs through the flax bushes and joins the boiling river, called Otumakorore, about twelve feet across and three feet deep.

This beautiful district is a perfect fairy scene, and a spot that one of these days will be selected by some enterprising member of the 'lost tribes' as a central sanatorium for restoring to health and re-invigorating jaded mortals from all parts of the world.

This hot river is perhaps the finest volume of clear hot water in the whole country; and magnificent baths could be constructed, more especially at a place where the cold stream joins the hot one, so that any temperature could be obtained, regulated at pleasure. After rest and refreshment, the tourist will leave his horses to feed on the luxuriant grass that abounds, and start on foot to explore this curious region. Crossing the bridge, he will continue his way along the path cut through the tall flax bushes, on leaving which he will turn sharp to the left up the course of the stream, and in the direction of some columns of steam that will be seen to be rising at the foot of the high range of hills which overlook the long valley. A short walk will bring him into the midst of a cluster of springs, which Hochstetter correctly describes as boiling wells. They are of different conformation to any others in the country, being literally like wells, eight or ten feet wide, and with the water reaching to within ten or twelve feet of the top. Some of these wells have in them clear water boiling furiously, and others thick water of a whitey-yellow colour, and others again nearly the ordinary lead-coloured mud, but all boiling. There are no silicious incrustations, but instead, the sides of the wells are clothed for some depth with thick vegetation of a purely tropical character; the beautiful club moss and delicate ferns growing down towards the boiling water, and actually living in the hot steaming atmosphere.

The Paeroa Plain.

The traveller will now return to the bridge, and resume the journey southward over a long flat of apparently good land, enclosed on the right by a series of low rounded hills covered with fern, and through a gap in which may be occasionally seen the high table-land Horohoro, and other rock-bound mountains and peaks. On the left, towering above the traveller, the Paeroa range rises considerably more than a thousand feet above the plain, and presenting an almost perpendicular face, covered in many places with heavy forests, while in others, wild, grand, rocky ribs or bare patches of burnt, fiery-looking earth are exposed. This level plain continues for more than fifteen miles, and for the entire distance the tourist rides immediately at the foot of the range, which rises sheer from the flat like a wall, without any preliminary broken ground. The Horohoro mountain range is of similar formation, but it is out of the line of internal fires and boiling fountains.

Many wonders are passed as we proceed, and soon after the warm river (Waikiti) is left behind, the Waiwhakahihi bush with its dark green foliage of magnificent Tobaras is conspicuous. Hochstetter admired this forest very much.

A little farther a scene of real picturesque beauty is revealed, which of itself is undoubtedly worth a visit. In the face of the range stands out a bare, grim rock, 800 feet in height, with a rough brown gigantic perpendicular front, broken and fissured into caves and crannies; wild, stern, and forbidding, it looks of itself like the castle of some dread ogre; but, investing it with a beauty not its own, a broad fringe of gigantic trees completely encircles it with a fanciful framework of the most graceful character. Nothing so truly picturesque is to be seen in the country, and its strange beauty is the more startling from the wild barrenness of the mountain face on which it is set. It is like a charming landscape of Millais hung on a bare brick wall. From the foot of the rock gushes a deliciously clear stream of ice-cold water.

Past this delightful picture volcanic indications are once more observed, and steam is to be seen rising from all portions of the range above and below, and soon a truly terrible scene is perceived. The range looks very thin, and from the further end of the saddle or neck rises a peak completely on fire. It is impossible to get at it, for the range here is burning in all directions, but seen from below it presents a dismal sight. Hochstetter says: 'The whole Paeroa range is full of wonders; numerous hot springs and mud-pools are boiling up at the foot, on the slope, and on the top of the range. The curious diseased-like red patches on the slopes, devoid of every trace of vegetation, point out from a distance the places from which sulphuric acid, sulphuretted hydrogen, sulphur, and steam are constantly escaping, producing fumaroles, hot springs, boiling mud-pools, and solfataras.'

The peak is devoid of any sign of vegetation, and is apparently of white clay, tumbled about roughly, as if shot out of a cart in large clods, and is literally on fire, for the sulphurous steam rises all over it, not in regular jets, but in intermittent clouds, like the steam from a rotten dunghill. Close to it are other broken peaks, whose bare sides of red, yellow, and white clay also give forth steam copiously.

There is no wilder place in the country than this, nor one where the subterranean fires appear to be much nearer the surface. At the foot of these hills are wonders without end, but the ground is so rotten and dangerous that the nervous had better content themselves with gazing at them from a respectful distance. One boiling pool here is very remarkable. Situated as it is close under the range, and the ground being so hollow and rotten, the crust sounds as if very thin. Close to it is another remarkable pool of boiling water or mud of a bright yellow colour, and several smaller ones of the same character are a little to the right. Between these and the track is the largest mud geyser in the whole Lake country. It is more than thirty feet across, and is decidedly a great sight, the blue-grey mud in it boiling furiously in great unctuous bubbles, which burst with a spluttering sound, and throw up irregular columns of mud. Many more smaller ones are in the vicinity, while around the boiling pools lie great masses of sulphur. Beyond this, steam rises in several places at different altitudes on the range,—five regular columns rising from a rough-looking place at the foot of the range, which proceed from little funnel-shaped holes. Farther still, right on the top of the mountain, is a mass of vapour equal in volume to that at Tikitere; it ascends from a large fountain called Kopiha, which no one has ever yet visited, and forms a conspicuous landmark for many miles. With the exception of a few jets emitted from bare red patches here and there on the mountain side, no more appearances of volcanic agencies are met with, and soon the long flax flat and the high perpendicular range are taken leave of, and a very broken fern country entered, a few miles of which bring the traveller into a gorge which opens on to the Waikato river, nearly a mile below the Orakei-korako Pa, which is on the opposite side, and distant from Ohinemutu about forty miles.

THE HOT SPRINGS OF ORAKEI-KORAKO.

There is a ferry here, and the tourist can camp on either side of the river he pleases. The scenery in this locality is eminently picturesque, the beautiful river winding its headlong course at the bottom of the valley, enclosed on both sides by high, irregular, partly bush-clothed ranges, which in some places rise sheer from the water, and in others recede a little in broken Manuka-covered mounds, or open out into deep gorges to admit the passage of some tributary, while here and there bays lined with delicate reeds or framed with a white sandy beach run into the banks, and furnish fine feeding grounds for numerous ducks, which may be seen quietly swimming about, while a solitary shag sits solemnly on a projecting branch or partly-submerged tree stump. If any disciple of 'Rob Roy' is in doubt which way to divert himself, he cannot do better than take his canoe to Taupo and descend the Waikato (over 200 miles) to its mouth. No end of scenes of wonder and of romantic and picturesque beauty will reward him, while the rapidity of the river is such, and the broken water so frequent, that plenty of excitement will add zest to the trip.

On the further side of the river, a walk of about a mile along the bank, in which numerous holes of boiling mud may be found, brings the traveller on to a small flat plateau of silica-incrusted rock about forty yards square, around which are numerous boiling pools, while little clusters of Whares are scattered about on the mounds around. Having completed his arrangements about board and lodging, the tourist will be at liberty to look around him, and he will admit a more extraordinary sight has not met his eyes since he left Ohinemutu.

Let him walk to the edge of the flat, rough terrace,

and stand on its brim immediately above the river, which runs furiously ten or twelve feet below. On the opposite side, at a distance of some forty or more yards, rises almost sheer from the water, to the height of from five to seven hundred feet, a range of rounded hills, for the most part covered with Manuka. Steam, as has already been described, rises in numerous jets from dozens of different localities in the Lake district; but if the tourist wishes to see the largest display of jets that is to be seen in the country, or most probably in the world, he must come to Orakei-korako. Hochstetter says: 'The chemical substances held in solution have coloured with every hue of the rainbow the rocks over which they flow, leaving the broad margin of pure white, which causes each waterfall to appear three or four times its natural size. The village takes its name from the great geyser at the foot of the hill, which raises a column of boiling water to a height of about forty feet. The funnel from which it rises is close to the bank of the river, in the midst of a place which looks like a great frozen snow-drift, surrounded by a number of very deep holes, down which water in violent action can be heard but not seen.'

The whole of the hills and bush visible from the crest where the village is built, are completely dotted with thousands of steam jets, whose little wreaths and clouds of steam keep coiling upwards from amongst the branches of the trees, giving a very singular character to an extremely beautiful landscape,—more particularly at early morning, when the steam jets are more clear and well defined. The view from the village, looking down upon the river, is very grand,— in swift course forming rapid after rapid, the Waikato goes plunging through a deep valley, between steep rising mountains, the water whirling and foaming round the rocky islands in the middle of the river, and dashing with loud uproar through the valley. Along the banks of the river white clouds of steam ascend from hot cascades, falling into it from basins full of boiling water, shut in by a white mass of stone. Sometimes only a few here and there play, then a whole lot burst out. To look down upon them for about half an hour or so, one cannot help thinking, with Hochstetter, that ' Nature is at work here, making experiments with a great scheme of water-works,— trying whether the fountains are all in perfect order, and the waterfalls have a sufficient supply, and the power employed calculated to keep the whole in working regularity.' He tried to count the places where a boiling water basin was visible, or where a cloud indicated the existence of such, and he counted seventy-six points, without being able to survey the whole region.

The scene over the river is really an extraordinary sight, for, independently of the marvellous numbers of steam columns, there are features of a most wonderful character altogether peculiar to the locality. Exactly opposite is a most remarkable formation of deposit, so altogether strange, that one can hardly believe it to be the sole handiwork of Nature.

Hanging down over the perpendicular bank, from a height of twenty or more feet, and nearly down to the water's edge, in festoons of the most charming character, is a cluster of stalactites, about thirty yards in width, looking like a section of an immense fringe. These stalactites are in the gaudiest of colours, dark reddish pink predominating, varied with copperas green and isolated streaks of white or bright orange yellow. The spring from which these deposits come is high up on the hill, and a broad steep terrace of the same colours winds up towards it, until lost to sight by a sharp turn into the tall Manuka. When on the other side, the tourist will examine this terrace, which is different in character as well as in colour from any in other localities. Further up the river are two or three smaller ones of the same nature, and also two or three small terraces of the purest snow-white. The scene altogether is unique. The narrow, rapid river furiously rushing along in heaving wavelets and clear shining billows,—the strange, brilliantly painted festoons hanging down over the water,—broad streaks of gaudy colours winding up the hill,—the glittering white terrace, so white as to be painful to look at,—and the countless jets of steam rising up above and below, to the right and to the left, on the bold, sombre-looking Manuka-clothed hill,—together form a picture not to be surpassed for strangeness and fascination probably in the world.

The side on which the tourist now stands is also by no means devoid of interest. At the left-hand river corner of the flat terrace on which he stands is a pot-like basin five feet across, in which is clear water gently boiling. Hochstetter says: 'The manner in which we practically experienced the intermittent properties of this fountain proves sufficiently how much caution is necessary in approaching such springs for the first time, and without experienced guides. My travelling companions wished to enjoy the luxury of a river bath early in the morning, and had just deposited their clothes, when suddenly they heard a violent detonation, and saw the water madly boiling up in a basin close by. They started back in afright, but only just in time to escape a shower-bath of boiling water, for now, amidst hissing and roaring, a steaming water column was being ejected from the basin in a slanting direction, and to the height of about twenty feet; the temperature of the water was 202° Fahr., and tasted like weak broth. These eruptions occur every two hours, and last about two minutes.' Of the terrace he says, 'Silicious, the recent sediment is soft as gelatine, gradually hardening into a triturable mass, sandy to the touch, and finally forming, by the layers deposited one above the other, a solid mass of rock of a very variable description at different places, both as to colour and structure. Here it is a radiated fibrous mass of light brown colour; there a chalcedony hard as steel, or a grey flint; at other places the deposit is white, with glossy conchoidal fracture like milk opal, or with earthy fracture like magnesite.'

This is what a scientific man sees: the ordinary observer only perceives a flat surface of rock apparently covered with a dirty white stucco-like substance. In and around it are several holes, mostly full of clear boiling water, but the main spring which formed it is at its head, close under the hill. Here there are two large basins, the water in which almost commingles. In the nearest one, however, the water is only pleasantly hot, and of a most extraordinary character, being to the touch as soft and silky as the finest and most delicate oil. It is said to be highly medicinal, and in it is a natural arm-chair of smooth marble-like substance, so that the bather can sit comfortably submerged nearly to his armpits.

The further basin is the main spring, and boils furiously. This spring must have altered its character considerably since Hochstetter visited it, for he makes no mention of its throwing out water, although he says that in 1848—the year of the great earthquake that nearly destroyed Wellington—it sent the water spouting to a height of 100 feet. As a matter of fact, it is the largest geyser in the whole country, and, with the exception of the caldron at the top of the White Terrace, and the Crow's Nest near Taupo, this throws out the largest body of boiling water (209° Fahr.). Every five minutes or so a jet is thrown out in a slanting direction to a distance of at least forty-five feet, while at the same time other smaller jets ascend perpendicularly, the display lasting each time for about half a minute, when it gradually subsides, though the boiling continues.

Just on the right of these, and sunk about a yard, is a pool into which the water of this spring trickles over a little terrace three feet high, similar in formation to the Pink Terrace. This pool is only about four feet across, but into it also runs an ice-cold stream, so that the bather can freeze or boil himself at pleasure, or can scald his feet and freeze his head, or *vice versa*, according to the dictates of his fancy.

In the flat down the river are several mud-springs, one of which is remarkable, being of that bright orange red often noticed on the margins of streams which contain iron; it boils steadily.

Re-crossing the river in a canoe at the same place already described, the tourist will climb the hill opposite by a path not very much out of the perpendicular, until a height of some three hundred feet above the river is attained, when he will follow its course up stream, along a path so choked with undergrowth that he has to force himself through it. More than once a dive is made down into a gully, so thickly overgrown with tall Manuka that the light of day is completely shut out, and the tourist splashes his way along in a gloom as of a dense forest.

Several mud-holes are passed, and many fragments of incrusted twigs lying about testify that at one time the whole hill was flowing with silicated water. Soon the ground becomes very rotten, and small holes emit steam in all directions. Suddenly a broad terrace is opened up nearly as wide as the Pink Terrace of Rotomahana, and across it runs a wall about ten feet in height, and festooned with white stalactites.

While walking up this terrace towards the wall, one soon becomes aware that its surface for a width of some ten or twelve feet from the edge is covered thickly with some very spongy, slimy substance of brilliant colours. This is the substance that forms the gaudy festoons already described, and is the deposit of a small stream which trickles along the edge of the terrace, and a little lower down spreads itself all over it. The water feels warm, and is of a very nauseous taste; the substance is very peculiar, and exactly resembles fungus. It is about two inches thick, and is composed of three layers or strata,— first, a thin layer of orange-pink, then a thin layer of bright green, and finally a thick layer of reddish-pink. Its nature from a scientific point of view is altogether unknown,—never having been analyzed,—but looks more of the vegetable than a mineral.

The geyser which formed the terrace was on another flat above the wall or step, and does not now play, so that the terrace is quite dry. This upper terrace is full of holes of boiling water, and sounds quite hollow, so that it is nervous work walking on it. It is covered with the same crystalline substance seen at the big conical geyser hills of Whaka-rewa-rewa. At the foot of the wall at the further end is a big caldron of clear blue water, boiling up intermittently to a height of about three feet, and making a great noise.

Having climbed the wall and crossed the upper terrace, another dive is made into the scrub, and another long rough scramble over some exceedingly rotten and dangerous ground has to be performed, when the tourist will find himself at the edge of what appears to be an old crater, for walls surround it on three sides, while a conical hill, burnt bare and red, and smoking, rises at each end. Nothing of particular interest is to be seen, though the appearance of the place is very wild and fascinating, but the native guide, passing cautiously over a very rotten patch of ground, suddenly disappears from view into the bowels of the earth.

The Alum Cave.

There is nothing to lead one to suppose that there is any underground district, the entrance being merely a well-like hole in the surface, completely overgrown with Manuka and other shrubs, but the tourist will follow his guide in mute surprise. Twelve steps, nearly perpendicular, and two feet in height and the same in width, will take him straight into another world. The character of his surroundings is entirely changed. He has been wandering about in the regions of Phlegethon, gazing on weird horrors that are unearthly in their grim aspects, supping his fill of wild dismal scenery and strange infernal marvels; and now the transformation scene is rung up, and in the twinkling of an eye, lo! he is in fairy-land, and stands staring around him in speechless astonishment, looking in vain for the scanty-skirted tinsel-decked sylphs that surely ought to be perched about on impossible flowers and moss banks. Tinne says: 'The walls and dome are incrusted with salts of various kinds and colours,—red, white, and green predominating; and the *tout ensemble* reminded me so strongly of the transformation scene in the pantomime of one's childhood, that I was disappointed to find no fairies there to welcome me to their retreat.' It is no use attempting to describe it, for nothing like it was ever seen out of the theatre.

It is no great cave, where one can wander about among corridors hung with festoons of stalactites, and jewelled with glittering crystals,—nothing of the sort. It is simply a big hole right into the centre of one of the conical hills already mentioned, formed by the earth falling into a hole burnt out by the volcanic fires. But it is so arranged, and the vegetation which has grown about is so appropriate, that as a work of fanciful art it could scarcely be improved on. At the foot of the steps the tourist finds himself at the wide mouth of a cave, which is entirely shut off from the outer world by a thick plantation of most magnificent tree-ferns, through which the sun struggles in glimmering rays, that dance about the interior of the cave like flashes from a lantern, glinting about among the alum-coated rocks and the blue pool of water as the winds play with the rustling feathery leaves of the palm-like ferns. The cave is about 200 feet to its furthest end, and about 150 feet high from the bottom to the edge of the roof of the entrance. Its width is about 45. It does not go straight into the hill, but down—both floor and roof—at an angle of about forty-five degrees. The floor is composed of large masses of rock, which near the bottom are covered with a white softish deposit of pure alum. At the bottom is a large pool of pleasantly-warm water, only a foot or two deep, and of a light blue colour, which tastes strongly of alum. The roof is of rock and reddish earth, and that portion of it just over the pool is incrusted with alum. Looking up from the bottom through the graceful ferns, the view is charming and fairy-like in the extreme.

This is the Alum Cave, the entrance to which is covered with a profusion of beautiful climbing plants,—the myrtle-leaved Rata (*Metrosideros robusta*), the Ponga (*Cyathea dealbata*) or silver tree-fern, and the Wheki or Tuakura (*Dicksonia squarrosa*),—their delicate tracery producing the most exquisite effects.

From Orakei-korako to Taupo is about twenty-eight or thirty miles, and, after climbing the hill above the Whares to the Pa on the top, a roughish ride brings the tourist to the foot of the range of bush-covered hills, and a path through fern and Tutu, which tower above the traveller, leads up it into a long flat among the fern hills, broken here and there with a gully or ravine. Fifteen miles of this brings the tourist on to the coach road, when a pleasant walk of eight or ten miles, through some rather pretty country dotted with clumps of bush, will take him to the summit of a range, from which a view of Lake Taupo suddenly reveals itself. Two or three miles before reaching Taupo, a fierce column of steam will be seen rising from the side of the hill on the left of the road. This is Karapiti, and is one of the most powerful steam-holes in the country, the force of the jet being so great that sticks and stones thrown in are immediately cast out again. It comes from a funnel-shaped hole of reddish earth. By climbing the rounded fern hill to the north, and following down the ridge, it is quite easy to reach. The ground between the fumarole and the gully running parallel to the road is very rotten, and must not be crossed. Steam issues in many places from its surface.

LAKE TAUPO AND TAPU-WAE-HARURU.

From the top of the hill above Taupo, the view is not unattractive. The lake, thirty miles across, is of itself exceedingly dreary in appearance, owing to the want of variety and entire absence of picturesque features on its shores; but if the air be clear, far away beyond its farthermost shore will be seen towering in majestic grandeur Mount Ruapehu, eternally snow-capped, and with its white mantle descending a third of its height. Twenty miles nearer, but looking close to it, is the conical volcanic Tongariro, from which vast columns of steam may be seen ascending.

Ruapehu is the highest mountain in New Zealand, except Mount Cook. Ruapehu summit is over ten thousand feet above the sea, while Mount Cook is thirteen thousand. Tongariro is nearly seven thousand, and is the only remaining volcano still retaining its active character, excepting, of course, that on White Island.

A mile or two to the left of the nearest point of the lake is the pretty rounded hill Tuahara, three thousand five hundred feet above the sea, and partly covered with bush; and at the margin of the nearest inlet of the lake is the white cluster of houses forming the village of Taupo (Tapu-wae-haruru), and a sharp canter down the slope will bring the traveller again to the Waikato a few hundred yards from its exit from the lake. It is very clear, and runs rapid, and is crossed by a strong and elegant bridge.

The village of Taupo is situated on a flat immediately overlooking the lake, and thirty or forty feet above it. There are hotels, stores, post office, telegraph office, a reading-room, and a cluster of private houses. The lake appears quite like a little sea, and the strong winds that sometimes blow across lash it into large billows, which break on the broad sandy beach with the true music of the sea-shore. A mile or two to the left, the water near the margin will be observed to be steaming for some distance, indicating that several subaqueous boiling springs exist near the shore.

The shores of the lake are uninteresting and dreary in the extreme, and the island of Motutaiko, near the centre, has no beauty to recommend it to the eye of the artist. But if the sky is cloudless, it will rest with lingering gaze on the rare beauty of the snow-clad king of mountains, — Ruapehu and its graceful attendant Tongariro. Strangely fascinating is the mighty mass, standing up clear and distinct, with its white mantle glinting with a brightness almost painful. So clear is the atmosphere, and so sharp the outline of the mountain, that it looks close at hand; and it is very difficult to realize that it is between sixty and seventy miles away. Its dark sister, Tongariro, is clothed in gloom, and, though twenty miles nearer, does not appear so distinct. On the right hand side of the main cone steam may be observed, and also on the smaller cone on the right. These two mountains redeem the scene from being entirely commonplace. From all the rest the eye turns away in weariness; but on these grand mountains it lingers long, especially if the tourist be up early enough to see the glory of the rising sun breaking in dazzling brightness over the broad expanse of glimmering snow.

The lake is one thousand two hundred and fifty feet above the sea, and the air is clear, dry, and bracing, so much so, that the cold, which of course is greater than on the coast, is not felt so much, and the district is very healthy; and only when the wind blows from the south it feels sharp, and bears on its airy atoms the cold biting breath of the distant South Island snow-fields.

In the neighbourhood of Tapu-wae-haruru are many things of interest to be seen, and the tourist will do well to spend a week inspecting them. The first day should be devoted to the bathing establishment, which is about two miles from the village. There is a good broad coach-road to it; the best way to reach it is by walking. When the road approaches nearest to the river, the pedestrian should turn off to the edge of the bank and inspect the scene, which is really quite charming. The river runs about one hundred and twenty feet below, between perpendicular cliffs, set off here and there with clumps of trees. It takes a sharp turn and runs with great rapidity, the gorge not being more than thirty yards in width, while the clear blue water, so clear that the cliffs can be seen reflected under the water for a great depth, is pleasingly contrasted with one or two green little islands. A little way down the river, a good many jets of steam will be seen to rise from the banks. The broad road gradually descends to the river, and trees have been planted on both sides, so that some day a splendid avenue will be formed. The block of land on which the springs are situated contains about fifteen thousand acres, and is the property of the Government, a European being placed in charge, whose establishment is in a gully running down to the river, and the perpendicular cliffs, fifty feet in height, which hedge the gully in, protect it from the cold south winds. It consists of a main building for sitting and eating, some detached cottages for sleeping in, a large covered-in bath, and numerous outhouses for horses, cattle, pigs, poultry, etc.

There are several vegetable gardens, a flower garden, and any quantity of fruit-trees scattered about. There are also three or four fine paddocks of sown grasses all enclosed.

There are three distinct baths of different temperatures, but all flowing out together. Two of them are twenty-five feet long by ten feet in width, and four feet deep. The water in one of these is quite hot, almost too hot to be pleasant, and in the other quite cold, each supplied by a copious stream of water. At the angle of conjunction these commingle; and a bath, of twenty-five feet in length and six in width by four in depth, is full of pleasantly warm water, slightly colder than above, the cold water being the denser. The water is kept in the bath by thick planks at the lower end, of such a height as to allow about six inches of water to flow over the top, so that the bather can climb over them, and set himself under a shower so formed. The whole is roofed in with rushes; and passion-flower, honeysuckle, clematis, convolvulus, perennial sweet-pea, Cape gooseberries, and other flowering creepers hang around in graceful and gaudily-coloured festoons. The dressing sheds are close at hand. Bathing in these baths is an enjoyment sensuous in the extreme. Many cures have been effected by them; and the water contains more mineral matter than that of any other spring yet analyzed. There is a large quantity of chlorine held, and also iodine, which no other spring contains.

A little distance to the right, and close to the road, is a mud-spring of the ordinary character. The broad road has been cut down to the river, and continued round in a circle so as to connect again half a mile above the bath. Near the river are some baths dug out alongside the hot stream, and a fall of a few feet gives a good shower.

Farther up the river, but not to be reached by following the bank, which is too steep, is a real wonder, and one of the most curious in the whole country.

The visitor is first shown what is known as the Witches' Caldron. This is a circular cavity in the side of the high river bank, shaped like an amphitheatre. Its wall is about twenty-five feet in height, and the hollow space about twenty feet across. The rocky sides are covered with deposits of brilliant colours,—red, crimson, green, orange, yellow, brown, and black being distinguishable. The bottom is a pool of blue water, boiling violently, close under the bank, and making considerable noise. Large volumes of steam rise from it, and it is very hot standing at the mouth of the cave. A few yards nearer the river, and within a couple of feet of the latter, is a long-shaped hole of beautifully clear water, also boiling furiously, and ejecting spray to the height of a foot or two. The river is very deep here, and deliciously transparent.

THE CROW'S NEST.

A hundred and fifty yards higher up, and close to the water, is the wonder above alluded to. It is called the Crow's Nest, and is a hollow cone six or seven feet in height and a dozen feet in diameter at the base, looking exactly like the roots and the portion of the trunk of a large tree roughly lined with twigs and branches, the whole being covered and petrified with mineral deposits. It is to these interior twigs that it owes its name, the similarity to a gigantic crow's nest being certainly very striking. At the bottom is a little pool of water, in which two holes can be seen far into the earth. As a rule it is quiescent; but every ten or fifteen minutes a mumbling noise is heard, when a cloud of steam immediately rises, followed by jets of spray, which ascend some eight or ten feet above the edge of the cove, and therefore about fifteen feet from the water. They are but trifling jets, but the Crow's Nest must nevertheless be looked upon as one of the true geysers.

On the top of the bank, and one or two hundred yards from the river, is a large patch of very rotten and burnt-up ground, in the centre of which is a hole called Big Ben, ten or fifteen feet deep, in which the boiling mud makes a loud regular noise exactly like the waste water of a screw steamer, ninety-three pulsations being made to a minute. Other holes are also in this dangerous bit of ground, which looks as though it would soon all fall in. The inspection of this will complete the objects of interest here.

The Huka Waterfall.

The Huka waterfalls should next be visited. These are about five miles from the village, along a rather rough path through the Manuka. The bridge over the Waikato is first crossed, and a path to the right taken, when a walk of some three miles brings the tourist again to the river valley. This valley is three or four hundred yards in width, and is enclosed by perpendicular cliffs fifty feet in height. At no distant date the river evidently filled up the whole of it, for the cliffs and the sides of numerous little conical and pyramidal peaks, once islands, are plainly watersheded, while the rocks lying about are scored by the action of gravel and water passing rapidly over them. The river is very pretty, and is about eighty yards in width, and runs swiftly. The rapids are soon reached. There is a narrow cleft in the rock, apparently opened by an earthquake, and two hundred and fifty yards long by an average of fifteen in width, and twenty-five feet deep, the bottom sloping off at an angle of about fifteen degrees. There are three ledges in the gorge, forming distinct rapids, and through it the water foams and seethes with tremendous fury, lashing itself into spray, on which the iridescent colours dance gaily. At the end of the gorge, the whole mass of the water takes a leap of twenty-five feet into a wide, calm, tree-lined basin, and then placidly continues its course. The fan-like fall, which looks hollow, but is in reality a solid mass of water, is very pretty; and the tourist can gaze his fill from the edge of the white solidified pumice rock immediately overlooking it. The best view, however, is from the bank about a hundred yards down the river, looking up over the calm basin to the white and foaming fall. The whole length of the rapids in the gorge can be examined from the edge of the rocks forming its sides, and it is truly fascinating to lie at full length, with the head over the edge, peering into the turmoil of water below. If one could imagine a fall of eighteen hundred yards in width and three hundred feet in height, the water of which, after falling, was forced through a gorge twice the width of Te Huka gorge, and three hundred feet in depth, he would have some notion of the extraordinary sight below the Victoria Falls of the Zambesi.

THE BITTER LAKE.

The next move of the tourist should be to visit Rotokawa. This lake is eight miles from Tapuwae-haruru, and to reach it a track is taken along the north-eastern side of Tuahara, which appears from that side as a small range, the centre peak being depressed into a crater. A small green pond in the bottom of a deep funnel-shaped hole, is passed on the way, which has been stocked with carp. The country beyond Tuahara stretches away in a fine flat sparsely covered with Tussoc grass, and here and there some Manuka scrub. In the distance may be seen the Tarawera mountain, and stretching away to the right a range of broken peaks and table-lands, dark with the hues of distance. After some miles of flat country, Rotokawa suddenly reveals itself in an extensive depression below. In visiting this lake the utmost caution has to be used, it being an extremely dangerous place, and no one should visit it alone. It is only a small lake of about a mile across one way by three-quarters the other. Descending to it the lake is skirted on its western side, and a spring of good fresh water being passed, the path leads through an expanse of burnt Manuka bushes, thrown to the ground as by a troop of elephants pushing over it. On the opposite side of the lake is a low range of some two or three hundred feet in height, continued for about half a mile beyond the head of the lake. Part of its sides have been shaken down by a recent earthquake, showing a bare surface of white pumice.

At right angles to the end of this range are a couple of rounded peaks connected by low undulating hills, and also bare in places, but showing bright red earth as at Rotomahana. In front of these peaks, and under the low range, steam is seen to rise in all directions. After walking a couple of hundred yards, the largest volume of steam is found to rise from a small lake about one hundred and twenty yards long by an average width of fifty, which is separated from the Rotokawa by a narrow neck of land covered with Manuku scrub, which neck, however, terminates within about a dozen yards of the main shore of the lake, the space between being hard rock, having on it a deposit of mud about an inch thick, which appears to be nearly all sulphur.

This rock sounds very hollow to the feet. The little lake boils up in hundreds of places, and, judging from the colour of some of the holes, must be exceedingly deep. In the centre are two boiling places, the water in and around which is a bright pea-green, other holes being bright blue, while the main body of water is of a dull whitey-blue, that in Rotokawa being a brownish green. The water of the latter has a nauseous sweet-acid taste. The crust round the small lake is extremely thin. In the side of the hill, in the low range already mentioned, one place will be particularly noticed from the brilliant colours it exhibits, pink and sulphur predominating.

From the head of the boiling lake to the rounded peaks is a wide open expanse of loose pumice and ash, sounding very hollow to the tread. This is cautiously crossed, and first a deep pool of boiling muddy water is seen, then a hollow flat where the pumice has fallen in is arrived at, and in this are many little mud-cones and hideous boiling mud-pools. To the left, over more very dangerous ashes, two sulphur mounds are met with, the sulphur lying about in blocks perfectly pure. Many other places are equally wonderful; altogether it is perhaps the most dismal, wild, and uncanny place in the entire Lake district, and the one in which the volcanic fires seem to have been most recently extinguished, if indeed they are extinguished, for the ash is in some places uncomfortably warm to the feet. The entire absence of life of any sort adds to the wildness of the scene, the weird, unearthly nature of which is such that it is quite a relief to get away from it, that the nerves may throw off the unnatural tension at which they have been strung.

Returning to the village, the tourist will now cross the lake by boat, or go along its eastern shore, passing numerous curious places, the traditions and anecdotes connected with which the natives will entertain him with till he reaches

TOKANO.

Tokano is a small bay extending southward from Pukawa; on the south side there is a magnificent waterfall of about 150 feet, where the Waihi stream falls over a bluff of rocks, and on the beach below hot water is bubbling out of the sand with a temperature of 153° Fahr. The natives have conducted this water into bathing pools with a temperature of 93° Fahr., and beautiful emerald-green ferns cover the places where the water flows. Hochstetter says: 'Silicious, not calcareous, sinter is deposited in them, but, strange to say, there is amid these alkaline springs also a chalybeate one of 156° Fahr., which deposits large quantities of iron ochre.' Some 500 feet above the lake, on the sides of the mountain, steam issues from innumerable places. The north side of the Kakaramea mountain seems to have been boiled soft and to be on the point of falling in; from every crack and cleft on that side of the mountain hot steam and boiling water is streaming out with a continual fizzing noise, as though hundreds of steam-engines were in motion. It was here that in 1846 the village Te Rapa was overwhelmed by an avalanche of mud, and the chief Te Heu Heu with his wives, for he had many, all perished. The fountains in and around Tokano it will be impossible to describe, they are so numerous, some of them throwing boiling water from 40 to 100 feet high, with temperature 186° to 208° Fahr. Hochstetter says: 'I believe if any one at Tokano or on the declivity of the Kakaramea would endeavour to count the several spots which give out either hot water, steam, or boiling mud, he would find more than five hundred of them.'

The tourist will now proceed southward towards Tongariro; he will pass Rotoaira, a lake three miles long, situated in a broad valley south of Pihanga. The foot of Tangariro is about twelve miles distant from this lake, which is 327 feet higher than Taupo. At the north-west end a swampy isthmus 100 yards broad joins a small peninsula to the main. This peninsula, called Motu-o-puhi, was formerly a place of refuge in times of war, and was strongly fortified. On the north-east shore is the native settlement of Tuku Tuku. This is a very pretty place, surrounded by magnificent forest trees, the ground gradually rising towards Pihanga mountain behind. On this rise there is a boiling spring, which has been for many years considered by the natives as an unfailing remedy for certain diseases, as they travel from a great distance to benefit from its healing qualities.

And now to reach the end of the journey the traveller will have to brace up and test his training for the ascent of the burning mountain.

The active volcano Tongariro, according to tradition, was but an empty crater until the arrival of the progenitors of the Maori race, which is supposed to have taken place early in the fourteenth century, when one of the ancestors of the Arawa, after landing on the east coast, started with a single slave to explore the country, leaving his wife at White Island to attend to the sacred fire which they had brought from Hawaiki. Having ascended Tongariro to survey the promised land, his slave fell ill from cold. Ngatoroirangi hailed his wife to bring him some of the sacred fire. She started forthwith, but scattered the fire by the way, and wherever any of the sparks fell springs began to boil, geysers burst forth through the fissures of the earth, and subterranean fires never more to be quenched produced the present fumaroles and solfataras. She arrived too late to save the life of the slave, so her lord threw the fire down the crater of Tongariro, where it has blazed and smouldered ever since. It still bears the name of the slave Ngauruhoe.

TONGARIRO VOLCANIC MOUNTAIN.

The first European who ascended Tongariro was Bidwell, in March 1839. From his *Rambles in New Zealand* we transcribe the account of how he fared on the sacred mountain :—' Leaving Rotoaira, the road leads over a tolerable level country covered with grasses of many different kinds, some of which would be well worth cultivating. As we skirted the base of the mountain in order to get the best place for the ascent, we found the ground marshy, and had to cross a great many small streams and nearly dry watercourses filled with large stones. About four o'clock we arrived at the junction of two considerable watercourses, and near some stunted trees. After being here for about half an hour, the clouds rolled out of the upper end of the valley, and we saw the cone was close to us. The Whanganui comes down here from the mountain, a noisy torrent about four feet deep. Here we camped for the night, and in the morning were astonished to find the mountain all round covered with snow except the cone, which was visible from the base to its apex, and appeared quite close. The natives said the mountain had been making a noise in the night; there seemed to be a little steam rising from the top, but the quantity was not sufficient to obscure the view. I set off immediately after breakfast, with only two natives, as all the others were afraid to go any nearer to the much-dreaded place, nor could I persuade the two who did set off with me to go within a mile of the base of the cone, where they made a fire and waited for my return. As there was no road, I went as straight towards the peak as I found possible, going over hills and through valleys, straight on. As I was toiling over a very steep hill, I heard a noise which caused me to look up, and I saw that the mountain was in a state of eruption, a thick column of black smoke rose up for some distance, and then spread out like a mushroom. As I was directly to windward I could see nothing more, and could not tell whether anything dropped from the cloud as it passed away. The noise, which was very loud, and not unlike that of the safety-valve of a steam-engine, lasted about half an hour and then ceased, after two or three sudden interruptions. The smoke continued to ascend for some time afterwards, but was less dense. I could see no fire, nor do I believe there was any, or that the eruption was anything more than hot water and steam, although, from the great density of the latter, it looked like very black smoke. I toiled on to the top of a hill, and was then much disappointed that the other side of it, instead of being like what I had ascended, was a precipice or very deep ravine, with a large stream of water at the bottom. With some difficulty I managed to get down, and on ascending the other side I found myself in a stream of lava perfectly undecomposed, but still old enough to have a few plants growing among the fissures. As I progressed towards the cone, which now seemed quite close, I arrived at another stream of lava, so fresh that there was not the slightest appearance of even a lichen on it, and it looked as if it had been ejected but yesterday; it was black, and very hard and compact, just like all the lava I have seen in this country; but the two streams were very insignificant, not longer at the utmost than three-quarters of a mile. I had no idea of 'a sea of rocks' until I crossed them. The edges of the stony billows were so sharp that it was very difficult to pass among them without cutting one's clothes into shreds. I at last arrived at the cone; it was, I suppose, of the ordinary steepness of such heaps of volcanic cinders, but much higher. I estimate it at 1500 feet from the hollow from which it appears to have sprung. It looks as if a vast amphitheatre had been hollowed out of the surrounding mountains in order to place it in. The sides of all the mountains around are quite perpendicular, and present a most magnificent scene. Thermometer at the base of the cone, in fine sunshine 65°, no shade to be had; barometer 25·14-20. The cone is entirely composed of loose cinders, and I was heartily tired of the exertion before I reached the top. Had it not been for the idea of standing where no man (at least European) ever stood before, I should certainly have given up the undertaking. A few patches of a most beautiful snow-white veronica, which I at first took for snow, were growing among the stones, but they ceased before I had ascended a third part of the way. A small grass reached a little higher; but both were so scarce that I do not think I saw a dozen plants of each in the whole ascent. After I had ascended about two-thirds of the way, I got into what appeared a watercourse, the solid rock of which, although presenting hardly any projecting points, was much easier to climb than the loose dust and ashes I had hitherto scrambled over. It was lucky for me another eruption did not take place while I was in it, or I should have been infallibly boiled to death, as I afterwards found that it led to the lowest part of the crater, and from indubitable proofs that a stream of hot mud and water had been running there during the time I saw the smoke from the top. The crater was the most terrific abyss I ever looked into or imagined; the rocks overhung it on all sides, and it was not possible to see above ten yards into it, from the quantity of steam which it was continually discharging. From the distance I measured along its edge I imagined it is at least a quarter of a mile in diameter, and very deep; the stones I threw in which I could hear strike the bottom did not do so in less than seven or eight seconds, but the greater part of them I could not hear. It was impossible to get on the inside of the crater, as all sides I saw were, if not quite precipitous, actually overhanging, so as to make it very disagreeable to look over them. The rocks on the top were covered with a whitish deposit from the steam, and there was plenty of sulphur in all directions, but the specimens were not handsome, being mixed with earth. I did not stay at the top so long as I could have wished, because I heard a strange noise coming out of the crater which I thought betokened another eruption. I saw several lakes and rivers, and the country about appeared covered with wood; the mountains in my immediate neighbourhood were all covered with snow. As I did not wish to see an eruption near enough to be either boiled or steamed to death, I made the best of my way down.'

The next person who ventured up the sacred mountain was Dyson, in March 1851. As his description is similar to Bidwell's, the mountain being quiet, we need not repeat it. Many tourists have attempted the ascent of Tongariro, but these two are the only Europeans who have ever gone up to the top of the burning cone.

WHITE ISLAND—BAY OF PLENTY.

White Island, or Whaka-ari, is an active volcano off the depth of the Bay of Plenty, 28 miles from the shore. It is about three miles in circumference, and 860 feet high. The base of the crater is $1\frac{1}{2}$ miles in circuit, and level with the sea. In the centre is a boiling spring, about a hundred yards in circumference, sending volumes of steam full 2000 feet high in calm weather. Round the edges of the crater are numerous geysers, sounding like so many high-pressure engines, and emitting steam with such velocity that a stone thrown into the vortex would immediately be shot into the air. Here and there are lakes of sulphureous water. The largest lake on the island is about 50 yards from the south shore, with a depth of two fathoms, the temperature 110° Fahr., the colour light green. There is a mud geyser on the south-east, on a slightly elevated bank, 12 feet in diameter, temperature 200° Fahr. The height of the lake above sea-level is 15 feet. There are no rocks of original formation. Vegetation is a dense scrubby green bush, and a short green grass. Area of the lake is 15 acres 3 roods 35 perches. The two highest peaks on the island on the east 845 feet, and that on the west 863 feet. From the edge of the crater the scene below is like a well-dressed meadow of gorgeous green, with meandering streams feeding the boiling caldron, but on approaching it is found to be the purest crystallized sulphur. No animal or insect lives on the island, scarcely a limpet on the stones; and 200 fathoms will hardly reach the bottom within half a mile of its shores. This island is the eastern limit of that extensive belt of subterranean fires which extends from Mount Egmont, through Tongariro, the Taupo and Rotomahana lakes, to Whale Island and the adjacent Ruarima rocks, north of which line earthquakes are rarely felt.

'ON THE INFLUENCE OF ATMOSPHERIC CHANGES ON THE HOT SPRINGS AND GEYSERS IN THE ROTORUA DISTRICT,' BY CAPTAIN G. MAIR.

'For many years past, partly from my own observations, and partly from conversations held with intelligent natives, I have been led to believe that some of the hot springs and geysers in the Rotorua and Taupo districts are affected to a remarkable degree by changes in the wind. Latterly, I have carefully noted down these changes, and will now give a few instances of this very remarkable phenomenon.

'Close to my residence at Tekautu, Ohinemutu, there is a large steaming pool 30 by 50 feet wide, and about 60 feet deep, named Tapui. It is situated on a grassy mound, about a hundred yards from Rotorua lake, and some fifteen or twenty feet above its ordinary level. I have been in the habit of bathing here for some years past, and generally found the water about blood heat.

'Since October 1874, I have observed that immediately the north and east winds (which blow directly across the lake) set in, Tapui (the hot water pool) fills up four or five feet, a strong outflow takes place, and the temperature rises from 100° to 190°. This continues till the wind shifts round to south, south-west, or west, when Tapui resumes its ordinary level and temperature.

'In 1875, from January to September, sea breezes or winds from north to east set in generally about 9.30, and at noon Tapui would be full and running over, and nearly at boiling point. In the evening, as the wind from the sea died away about six o'clock, the water began to recede, the temperature to lower, and at eight o'clock the water became cool enough for bathing.

'This year, however, the prevailing winds have continued to blow from the sea, and Tapui has seldom been fit to bathe in. For many years the natives living at Koutu have observed the rise and fall of this spring, which circumstance has been passed into a proverb—"Tapui tohu hau" ("Tapui the wind pointer"). They tell me that they never have known it to remain hot for so long a time previously.

'At Whaka-rewa-rewa, two miles and three-quarters from Ohinemutu, there are several hundred mud-baths and boiling springs. There are also several fine geysers, which become very active during south-west and westerly winds, frequently throwing water 40 to 60 feet. The principal ones are named Pohutu and Te Horu. They are rarely active in the middle of the day, but generally between seven and nine in the morning, and from three to five in the evening, while Whakaha-rua, or the "Bashful Geyser," is only in a state of violent ebullition after dark.

'Perhaps the most singular instance of atmospheric influence is in the case of Te Tarata, the White Terrace, at Rotomahana. The great crater, which is about 90 feet in diameter, is usually full of deep azure blue coloured water, occasionally boiling up 10 or 15 feet, but when the keen south wind or Tonga blows, the water recedes, and you can descend 30 feet into the beautifully incrusted crater, which remains empty till the wind changes, when it commences to refill at the rate of three or four feet per hour, boiling and roaring like a mighty engine. When the crater is almost full, grand snow-white columns of water, 20 feet in diameter, are hurled 60 feet into the air. Blue waves of boiling water surge over the shell-like lips of the crater, and fall in a thousand cascades over the alabaster terraces.

'There are many other springs (for example, Ohaki near Taupo, Whaka-poa-poa at Orakei-koraka) which, according to Maori legends, are influenced by changes in the wind. There is a great spring called Ketetahi, situated on the western slope of Tongariro, and 1800 feet above the level of Rotoaira lake, which is only active during westerly winds.

'About three miles north of the Waikato river, at Niho-o-te-kiore, and in the middle of Hinemara plain, are two fine springs named Waimahana. These pools are circular, each about 25 feet in diameter, and 30 or 40 yards apart. They are situated on a spur which slopes down to the Whangapua river 180 feet below, on the sides of which the outflow has formed pretty white silica terraces. The northernmost pool slowly bubbles, and the temperature throughout the year ranges from 190° to 200°. In March or April the water in the other pool recedes to 10 or 15 feet below the surface, and remains at blood heat until December, when it fills up; a strong outflow takes place, and the temperature is increased to 204°.

'I carefully noted these springs during the years 1870 to 1874 without detecting any deviation from what I have already stated.'

ANALYSIS OF WATER FROM A FEW OF THE MEDICINAL SPRINGS IN THE HOT LAKE DISTRICT, BY THE GOVERNMENT ANALYST.

Manupirua Soda Spring, on the south-east shore of Rotoiti, is a beautifully clear pool, 20 feet in diameter, having a temperature of 107° to 110° Fahr., at the foot of a high pumice cliff, on the immediate shore of the lake. The water is clear, with a bluish tinge, harsh to the touch, and deposits sulphur. This pool has a strong outflow of 40 to 50 gallons per minute, and is reputed to have great curative properties. It is better known as the 'Tikitere' medicinal spring. Analysis—mono-silicate of lime, 1·51 ; of iron, ·99 ; of manganese, ·77 ; sulphate of soda, 11·50; of lime, 2·43 ; chloride of potassium, ·47 ; of sodium, 6·25 ; silica, uncombined, 8·53, with traces of lithium, iodine, and sulphuretted hydrogen : total, 32·45 grains to the gallon.

Kuirua Washing Fountain is in the native village of Ohinemutu, where a strong stream flows from a number of hot springs which cover an extent of about 30 acres. This one has a temperature of from 136° to 156°, and is so soft that clothes can be washed—and nearly all the clothes of the village are washed—in it without the use of soap. It deposits a white flocculent sediment in the bottles, leaving the water clear, with a faint yellow tint and an alkaline reaction. Analysis—mono-silicate of soda, 2·57; of lime, ·34; of magnesia, ·12; of iron, ·31; sulphate of soda, 10·31; chloride of potassium, 2·08; of sodium, 45·70; silica, free, 18·42; and traces of phosphate of alumina: total, 79·85 grains to the gallon.

Ta-Pua Te-Koutu is a large pool, 60 to 80 feet deep, about three-quarters of a mile from Ohinemutu. The temperature of the water in this pool is from 90° to 100° Fahr. with westerly or southerly winds; but if the wind change to north or east, the water rises four feet in level, and the temperature increases to 180° Fahr., with a strong outflow. Thick masses of slimy confervoid plants live in the bottom of the pool. The water in the bottles was clear and colourless, with an alkaline reaction. Analysis—silicate of soda, 32·12; mono-silicate of lime, 1·61; of magnesia, ·40; of iron, ·67; sulphate of soda, 7·06; chloride of potassium, ·97; of sodium, 29·94; and traces of phosphate of alumina, iodine, and lithia: total, 82·78 grains to the gallon.

Te Kau Whanga Doctor at Sulphur Point, $1\frac{1}{4}$ miles from Ohinemutu, is a powerful sulphur bath, having a temperature of 204° Fahr. The water as received was clear and colourless, with a distinct acid reaction, and evolving an offensive odour. It deposited a brownish sediment on being boiled. This bath is reported to have great curative properties, and is known to tourists as the 'Pain Killer.' Numerous marvellous cures of long-standing diseases are attributed to this spring. Analysis—sulphate of potash, 2·96; of soda, 34·37; chloride of sodium, 59·16; of calcium, 3·33; of magnesia, 1·27; of iron, ·25; silica, 16·09; hydrochloric acid, 7·60; sulphuretted hydrogen, 2·01; and traces of phosphate of alumina, lithium, and iodine: total, 127·04.

Sulphur Bay Spring, on the edge of Lake Rotorua, is formed by innumerable small jets, forced up through sand, having a disagreeable odour, and a temperature from 90° to 100° Fahr. This bath is reported to have a powerful action on the skin, owing no doubt to the large quantity of sulphuric acid it contains. As received it was colourless, with a slight flaky sediment. Analysis—sulphate of potash, 0·7; of soda, 8·37; of lime, 2·50; of magnesia, ·93; of iron, 2·68; sulphuric acid, free, 18·02; hydrochloric acid, free, ·86; silica, 10·08; sulphuretted hydrogen, 1·01; and traces of sulphate and phosphate of alumina: total, 44·52 grains to the gallon.

Te Kau Whanga Mud Bath, at Sulphur Point, is a thick brown muddy water, covered with an oily slime, and having a temperature of 80° to 100° Fahr. When received it had deposited a heavy muddy sediment, and had a persistent acid reaction and an offensive odour. Analysis—sulphate of potash, ·77; of soda, 23·71; of alumina, 1·46; of lime, 2·04; of magnesia, 1·62; of iron, 1·47; sulphuric acid, 7·60; hydrochloric

Analyses of Medicinal Springs. 165

acid, free, 7·66; sulphuretted hydrogen, 3·19; silica, 13·86; and traces of phosphate of alumina, iodine, and lithium: total, 63·38 grains to the gallon.

Perekari Aperient is a boiling pool in the sandpit at Sulphur Point, and near the edge of the lake; temperature of the water, 130° to 150° Fahr. The water is discoloured, and has a very offensive smell. As received it was clear and colourless, with a strong acid reaction; it had deposited a great deal of sediment, which consists of nearly pure silica. Analysis —sulphate of soda, 26·75; of lime, 2·45; of magnesia, 1·86; of iron, ·76; chloride of potassium, ·63; hydrochloric acid, free, 5·38; silica, 18·17; and traces of sulphate and phosphate of alumina and lithia: total, 56·00 grains to the gallon.

Ariki Kapakapa, at Whaka-rewa-rewa, is a small pool with a strong outflow, having a temperature of 160° Fahr. It deposits sulphur, and is surrounded by a great number of other baths and mud volcanoes. It is reported to have powerful curative properties. It was colourless as received, with a heavy deposit of silica, and an acid reaction which was persistent at its boiling point. Analysis—sulphate of potash, ·38; of soda, 12·51; of alumina, ·68; of lime, 1·21; of magnesia, 1·29; of iron, 3·15; sulphuric acid, free, 13·95; hydrochloric acid, free, 2·62; silica, 18·94; and traces of phosphate of alumina, iodine, and lithium: total, 54·94 grains per gallon.

Turi Kore Waterfall, at Whaka-rewa-rewa. The

sample was taken from the waterfall, which drains from a large pond 300 yards long, the reservoir of a number of boiling springs that are in continual activity. The temperature of the fall is from 96° to 120° Fahr. The water is of a dirty brown colour, and is in great repute among the Maoris for the cure of all cutaneous diseases. Analysis—silicate of soda, 16·32; of lime, 1·61; of magnesia, 1·14; of iron, ·39; sulphate of soda, 13·47; chloride of potassium, 1·24; of sodium, and traces of phosphate of alumina, lithium, and iodine: total, 87·78 grains to the gallon.

Koroteoteo Oil Bath, at Whaka-rewa-rewa, is a strong boiling stream, the recorded temperature being 214° Fahr., from two springs, one of which, surrounded by beautiful sulphur incrustations, throws a powerful jet to a height of twenty feet. The water is distinctly alkaline, or slightly caustic, which is probably the reason of its being termed an oil bath. Analysis—mono-silicate of soda, 2·08; of lime, 3·16; of magnesia, ·76; of iron, ·85; sulphate of soda, 7·49; chloride of potassium, 1·46; of sodium, 66·34; silica, free, 22·40; and traces of chloride of lithium, phosphate of alumina, and iodine: total, 104·54 grains to the gallon.

Te Kute Great Spring, on the road to Paeroa, 10½ miles from Ohinemutu, is a pool three-quarters of an acre in extent, the temperature varying from 100° to 212° Fahr. in various parts. It boils furiously, and dense volumes of steam are constantly rising from it. The water is of a muddy brown colour, and contains a large proportion of sulphuretted hydrogen, and is reported to be wonderfully efficacious in cases of rheumatism and cutaneous diseases. Analysis—sulphate of potash, ·59; of soda, 12·66; of alumina, 11·22; of lime, 1·01; of magnesia, ·69; of iron, 1·73; sulphuric acid, free, ·77; hydrochloric acid, free, 1·63; sulphuretted hydrogen, 5·74; silica, 12·40; and traces of phosphoric acid and iodine: total, 484·4 grains to the gallon.

Te Mimi Okakahi, a waterfall, having a temperature of 90° to 112° Fahr. It drains from Te Kute, the spring mentioned above, and only differs from it in being more dilute, and having a larger proportion of sulphuric acid, and less sulphuretted hydrogen. Analysis—sulphate of potash, ·13; of soda, 4·78; of lime, 2·04; of magnesia, ·93; of iron, ·23; sulphuric acid, 12·48; hydrochloric acid, free, 3·82; sulphuretted hydrogen, ·98; silica, 4·12; and traces of sulphate of alumina, phosphate of ammonia, and iodine: total, 29·51 grains per gallon.

Te Tarata, the spring which forms the great White Terrace of Rotomahana. This is a true geyser, having a large crater-shaped basin, 90 feet in diameter, the lip of which is about 70 feet above the level of the lake. This basin is emptied by an explosive effort, which throws the water to a height of 40 feet, emptying the basin, which again fills up rapidly. The water trickles over the ledges of the terraces, depositing fresh layers of silicious sinter as it cools in its passage to the lake. The water in the basin has a deep azure blue colour, and a temperature of 210° Fahr. Analysis — silicate of soda, 68·48; mono-silicate of lime, 1·62; of magnesia, ·53; of iron, ·51; sulphate of soda, 7·84; chloride of potassium, 2·87; of sodium, 62·61; and traces of phosphate of alumina and lithia. All but soda are mono-silicates; the little excess of silica, 7·16, is included in the soda silicate. Total, 144·46 grains to the gallon.

The Acid Bath, at the bottom of the Pink Terrace, and close to the lake, is a muddy pool 20 feet in diameter, having a temperature of 109° to 115° Fahr. It is kept in a state of ebullition by the powerful escape of gas, which causes faintness when inhaled. The pool has no outflow, and the water is of a dirty chocolate colour. As received, the water had a persistent acid reaction and offensive odour; it had deposited a silicious sediment in large quantities. Analysis—sulphate of potash, ·94; of soda, 33·47; of lime, 2·11; of magnesia, 1·14; of iron, 1·20; sulphuric acid, free, 76·79; hydrochloric acid, free, 7·28; sulphuretted hydrogen, ·41; silica, 7·01; and traces of sulphate and phosphate of alumina and lithia: total, 130·35 grains to the gallon.

Otukapuarangi, the Pink Terrace of Rotomahana. This terrace has been built up round a great circular pool, 180 feet in diameter, from which there is a strong outflow of clear bright water, having a temperature of 204° to 208° Fahr., and depositing silicious sinter, of a delicate pink tint, in large quantities. As received, the water was faintly acid, changing to alkaline when boiled. Analysis—silicate of lime, 1·91; of magnesia, 1·16; chloride of potassium, 1·05; of sodium, 93·55; sulphate of lime, 10·96; of soda, 1·01; alumina as phosphate, ·54; silica, free, 43·95; and traces of iron oxides and lithia: total, 154·13 grains to the gallon.

The presence of iodine and bromine are in small quantities, as also the metal lithium; but as this substance has active medical properties, even when administered in small quantities, if continuously, it is often an important matter that its presence in any mineral water should be known to those who use it. These waters will be found quite as useful as those alkaline waters of the European spas, in which the alkalies are combined with carbonic acid, and where neither iodine nor lithium are present to any notable extent.

Some of the acidulous waters have useful medicinal qualities to a remarkable degree, and should prove efficacious in cases of rheumatism and skin diseases.

It will be seen from the foregoing that we have several kinds of mineral waters here, both cold and hot, within two days' easy drive from Auckland; and although the springs can be counted by the hundred, they are at no great distance from each other, a circumstance likely to be of considerable advantage to many who may desire to use mineral water for their health.

It is also important to observe that, while there is this difference in the constitution of their saline constituents, they nearly all contain iodine in sufficient amount to impart to them very decided therapeutic properties; this substance, it may be stated, has been proved to be very efficacious when externally applied in cases of cutaneous eruptions.

In some of the springs, the analyst found a large quantity of free (native) iodine, as much as 2·17 grains to the gallon; the water resembling those of Wiesbaden, Kreutznach, and Aix-la-Chapelle of the Continental spas, and those of Cheltenham, Harrogate, and Leamington of the English ones; with a superior strength over the English ones, and therefore more important medical effects may be expected.

On the whole, it is hardly possible to over-estimate the valuable curative properties of these springs, and the wide range in their character.

INDEX.

	PAGE
Alum Cave at Orakei-korako,	133
Analysis of Water from Fifteen Springs,	162
Auckland to Tauranga by Sea,	28
Auckland by the Waikato (Rail and Coach) to the Hot Lakes,	19
Battle of Rangiriri, Waikato War,	24
Bidwell's Ascent of Tongariro Burning Mountain,	151
Bitter or Alum Lake, The,	146
Captain Cook's Observations,	29
Crow's Nest, Waikato River,	143
Gate Pa, the Story of the Massacre,	36
Great Nga-hapu Boiling Fountain,	99
Hinemoa, Maori Love Story,	59
Huka Waterfall, Waikato River,	144
Kapiti Fountain,	104
Koheroa, the First Battle in the Waikato,	22
Kuakiwi Springs,	103
Mercury Bay and Captain Cook,	29
Ohinemutu Native Village and Springs,	51
Ohinemutu to Orakei-korako,	118
Ohinemutu to Wairoa,	78
On the Influence of Atmospheric Changes,	158
Orakei-korako Native Village and Springs,	124
Orakei-korako to Tapu-waeharuru,	137
Oropi Eighteen-mile Bush,	42
Paeroa Range of Burning Mountains,	118
Pink Terrace, Rotomahana,	111
Rangipakaru Fountain,	107
Rotokakahi Lake,	79
Rotomahana Hot Lake,	85
Rotomahana, Eastern Side of the Lake,	99
Roto Pounamu' or Green Lake,	105
Rotorua Lake,	49
Steam Fog-horn at Rotomahana,	101

	PAGE		PAGE
Sulphur Point, Rotorua,	67	Tokano Native Village and Springs,	149
Sunset at Rotomahana,	110	Tongariro Burning Mountain,	151
Takapo Fountain,	102	Wai-kanapanapa Springs,	102
Tarawera Lake,	80	Whaka-rewa-rewa Springs,	69
Tauranga to Maketu, Bay of Plenty,	45	White Island Volcano, Bay of Plenty,	156
Tauranga to Ohinemutu,	34	White Terrace, Te Tarata, Rotomahana,	89
Tauranga Township,	33		
Tikitapu Lake,	78		
Tikitere Sulphur Springs,	47		

www.ingramcontent.com/pod-product-compliance
Lightning Source LLC
Chambersburg PA
CBHW020308170426
43202CB00008B/537